普通高等教育"十一五"国家级规划

机械工业出版社精品教材

模具设计与制造专业英语

A NEW ENGLISH BOOK FOR DIE OR MOLD DESIGN AND MANUFACTURING

第 3 版

Third Edition

主　编　王晓江
参　编　钱泉森　卢端敏　林　涵
　　　　吴　斌　陈永兴　孙　慧
主　审　王兆奇

机 械 工 业 出 版 社

本书内容均选自英、美等国专业教材及专业刊物中的原文，共 10 个单元、30 课，90 余篇（其中课文 30 篇，阅读材料 60 多篇）。内容涉及模具材料及热处理、机械制图与公差配合、刀具和夹具设计、冲压和塑料成型机械、冲压工艺及模具设计、塑料及塑料模具设计、压铸模具和锻模设计、常用机械加工方法、特种加工工艺、模具 CAD/CAM、数控加工技术、快速成型与先进制造技术等方面。内容全面，难易适中，图文并茂，配套有课文参考译文，便于阅读和学习。

本书为高职高专模具设计与制造专业教材，同时也可作为从事模具设计与制造的企业工程技术人员自学参考书。

本书配套有电子课件，凡选用本书作为教材的教师可登录机械工业出版社教育服务网 www.cmpedu.com，注册后免费下载。咨询电话：010-88379375。

图书在版编目（CIP）数据

模具设计与制造专业英语/王晓江主编 . —3 版 . —北京：机械工业出版社，2018.6（2023.1 重印）

普通高等教育"十一五"国家级规划教材修订版 机械工业出版社精品教材

ISBN 978-7-111-60120-3

Ⅰ. ①模… Ⅱ. ①王… Ⅲ. ①模具—设计—英语—高等学校—教材 ②模具—制造—英语—高等学校—教材 Ⅳ. ①TG76

中国版本图书馆 CIP 数据核字（2018）第 115530 号

机械工业出版社（北京市百万庄大街 22 号 邮政编码 100037）
策划编辑：于奇慧 责任编辑：于奇慧
责任校对：刘 岚 封面设计：马精明
责任印制：李 昂
北京中科印刷有限公司印刷
2023 年 1 月第 3 版第 4 次印刷
184mm×260mm · 13.5 印张 · 324 千字
标准书号：ISBN 978-7-111-60120-3
定价：39.80 元

电话服务 网络服务
客服电话：010-88361066 机 工 官 网：www.cmpbook.com
　　　　　010-88379833 机 工 官 博：weibo.com/cmp1952
　　　　　010-68326294 金 书 网：www.golden-book.com
封底无防伪标均为盗版 机工教育服务网：www.cmpedu.com

第3版前言

本书为全国机械职业教育模具类专业教学指导委员会规划教材、普通高等教育"十一五"国家级规划教材修订版。自2008年第2版出版以来，先后重印了11次，受到许多院校师生的欢迎。本次修订是在第2版的基础上，依据部分使用本书的院校师生和模具企业工程技术人员的意见和建议，对教材编排顺序、内容做了适当的修订。主要有以下几个方面：

1) 对原来的课文次序进行了重新调整和编排。

2) 对原教材部分课后阅读材料内容进行了适当的删减。

3) 在保持原教材内容和体系的基础上，增编了 Lesson 12 、Lesson 13 两课内容。

4) 适当添加了部分示图、专业术语等。

5) 本次修订还增加了 Appendix B   Tables of Weights and Measures。

修订后的教材分10个单元、共30课。建议60学时完成。各校在组织教学时，可根据本校和学生的具体情况，不受教材编排顺序和内容限制，进行适当的删减和调整。

本书在修订过程中得到了机械工业出版社的大力支持和帮助，陕西工业职业技术学院肖春艳、杨燕老师对教材提出了宝贵的修改建议，谨向他们表示衷心的感谢。

由于编者水平所限，书中难免有不少缺点和错误，敬请读者批评指正。

编　者

第2版前言

本书是普通高等教育"十一五"国家级规划教材、全国机械职业教育模具设计与制造专业教学指导委员会规划教材。

本书第1版自2001年出版以来，先后重印10余次，深受广大高职院校师生的欢迎和好评。此次修订，是在第1版的基础上，依据模具设计与制造技术发展对高技能人才的需求，征求了部分院校师生的意见，并听取了模具企业工程技术人员的建议，对原书内容做了适当的更新和扩充。主要进行了以下几个方面的修订工作：

1. 在保持第1版教材内容和结构体系的基础上，新增了2个单元共4篇有关机械制图、第三角画法、公差配合、先进制造技术、快速成形技术等内容的文章（第25～28课），同时还补充了部分课后阅读材料。

2. 为方便读者学习，增加了第1～28课课文的参考译文。

3. 对第1版教材的部分课后阅读材料做了适当的删减。

修订后的教材共12个单元28课。各校在组织教学时，可根据本校学生的实际情况进行适当的删减和调整，不必受教材编排顺序和内容的限制。

王晓江任本书主编，选编并翻译了第25～28课课文，史铁樑翻译了第16～24课课文，陈永兴翻译了第5～10课课文，南欢、董海东、孙慧、王建军、周勋、杨新华等翻译了其他课文。

本书在编写过程中，得到了王明哲、刘航、殷铖、胡占军等老师的大力支持，在此谨向他们表示衷心的感谢。刘宝兴老师对教材提出了许多宝贵的修改意见，在此深表感谢。

由于编者水平所限，书中难免有不少缺点和错误，敬请读者批评指正。

编　者

第1版前言

本书是根据教育部"关于加强高职高专教育教材建设的若干意见"和国家机械工业局教材编辑室"关于组织新编高职高专模具专业教材的原则"以及"模具设计与制造专业英语"课程教学大纲编写的。

其目的是为了更好地帮助模具专业学生进一步适应国际、国内模具专业发展的需要，提高直接阅读原文和翻译有关专业英语书刊的能力，学习和借鉴国外先进的模具设计和制造技术，特别是模具 CAD/CAM 应用软件等，从而大力推进我国模具行业和模具产品的快速发展。

本书内容均选自英、美等国专业教材及专业刊物中的原文，共 80 篇（其中课文 25 篇，阅读材料 55 篇），分 10 个单元，24 课。内容比较全面，涉及模具材料及热处理、刀具和夹具、冲压和塑压成型机械、冲压工艺及模具设计、塑料及塑料模具设计、压铸模具及锻模设计、常用冷加工方法、特种加工工艺、模具 CAD/CAM 和数控加工技术等方面。

本书可供高职高专模具设计与制造专业学生作为教材使用，教学中可根据各校的实际情况，调整授课顺序或删减有关内容。同时也可供有关模具设计与制造企业的工程技术人员参考。

本书由陕西工业职业技术学院王晓江主编，参加编写的人员有无锡工业职业技术学院吴斌、福建职业技术学院林涵、江西省机械工业学校钱泉森、成都市工业学校史铁樑、张家界航空工业学校卢端敏。在编写过程中得到了夏克坚、赵居礼、戴勇、翁其金等同志的大力支持。

本书由陕西工业职业技术学院王兆奇主审，陕西工业职业技术学院澳大利亚籍教师 Jared Morise 审阅了全书，参加审稿的有刘全胜、朱燕清、徐政坤、彭雁、包杰、武友德、陈勇、刘航等老师。他们对本书提出了许多宝贵的修改意见，在此表示衷心的感谢。

由于编者水平有限，加上时间紧迫，经验不足，书中难免会有缺点和错误，欢迎读者批评指正。

编　者

CONTENTS

Unit One

Lesson 1　Tool Materials

Text

The specific material selected for a particular tool is normally determined by the mechanical properties necessary for the proper operation of the tool. These materials should be selected only after a careful study and *evaluation* of the function and requirements of the proposed tool. In most applications, more than one type of material will be *satisfactory*, and a final choice will normally be governed by material *availability* and economic considerations.

The principal materials used for tools can be divided into three major *categories*: ferrous materials, nonferrous materials, and nonmetallic materials. Ferrous tool materials have iron as a base metal and include tool steel, alloy steel, carbon steel, and cast iron. Nonferrous materials have a base metal other than iron and include aluminum, magnesium, zinc, lead, bismuth, copper, and a variety of alloys. Nonmetallic materials are those materials such as woods, plastics, rubbers, *epoxy* resins, ceramics, and diamonds that do not have a metallic base. To properly select a tool material, there are several physical and mechanical properties you should understand to determine how the materials you select will affect the function and operation of the tool[1].

Physical and mechanical properties are those characteristics of a material which control how the material will react under certain conditions. Physical properties are those properties which are natural in the material and cannot be permanently altered without changing the material itself. These properties include: weight, color, *thermal* and electrical conductivity, rate of thermal expansion, and melting point. The mechanical properties of a material are those properties which can be permanently altered by *thermal* or mechanical treatment. These properties include strength, hardness, wear resistance, *toughness*, *brittleness*, plasticity, ductility, *malleability*, and *modulus* of elasticity.

From a use *standpoint*, tool steels are utilized in working and shaping basic materials such as metals, plastics, and wood into desired forms. From a composition standpoint, tool steels are carbon alloy steels which are capable of being hardened and tempered. Some desirable properties of tool steels are high wear resistance and hardness, good heat resistance, and *sufficient strength* to work the materials. In some cases, *dimensional stability* may be very important. Tool steels also must be economical to use and be capable of being formed or machined into the desired shape for the tool.

Since the property requirements are so special, tool steels are usually melted in electric furnaces using careful *metallurgical* quality control. A great effort is made to keep *porosity*, *segregation*, *impurities*, and nonmetallic inclusions to as low a level as possible[2]. Tool steels are subjected to careful *macroscopic* and *microscopic inspections* to ensure that they meet strict "tool steel" specifications.

Although tool steels are a relatively small percentage of total steel production, they have a strategic position in that they are used in the production of other steel products and engineering materials. Some applications of tool steels include drills, deepdrawing dies, shear *blades*, punches, extrusion dies, and cutting tools.

For some applications, especially where extremely high-speed cutting is important, other tool materials such as sintered *carbide* products are a more economical alternative to tool steels. The exceptional tool performance of sintered carbides results from their very high hardness and high *compressive* strength. Other tool materials are being used more and more often industrially.

Questions

1. What is meant by the term "physical properties of a material"?
2. What are the mechanical properties of a material?
3. What makes a material either ferrous or nonferrous?
4. What are some applications of tool steels?

New Words and Expressions

1. evaluation/iˌvælju'eiʃən/n. 评价，估价，鉴定，计算
2. satisfactory/sætis'fæktəri/a. 满意的，符合要求的
3. availability/əˌveilə'biliti/n. 存在，具备，有效性，利用率
4. category/'kætigəri/n. 种类，范畴，类型
5. epoxy/e'pɔksi/a. 环氧的
6. thermal/'θəːməl/a. 热（量）的，由热造成的；n. 上升暖气流
7. toughness/'tʌfnis/n. 韧性，韧度，塑性，刚度
8. brittleness/'britlnis/n. 脆性，脆度，脆弱性
9. malleability/mæliə'biliti/n. 可锻性，延展性，展性
10. modulus/'mɔdjuləs/n. 模量，模数，系数
11. standpoint/'stændpɔint/n. 观点，立场

12. sufficient/sə'fiʃənt/a. 充分的，足够的
13. strength/streŋθ/n. 力（量），强度
14. dimensional/di'menʃənl/a. 尺寸的，量纲的
15. stability/stə'biliti/n. 稳定（性），安定度
16. metallurgical/metə'lɔːdʒikəl/a. 冶金（学）的
17. porosity/pɔː'rɔsiti/n. 多孔（性），孔隙度，疏松（度）
18. segregation/segri'geiʃən/n. 分离，分开，隔离，偏析
19. impurity/im'pjuəriti/n. 杂质
20. macroscopic/mækrə(u)'skɔpik/a. 宏观的，肉眼可见的
21. microscopic/maikrə'skɔpik/a. 显微镜的，微观的，微小的
22. inspection/in'spekʃən/n. 检查，调查，

参观，视察

压缩力的

23. blade/bleid/*n.* 刀口，刀片，刀身

24. carbide/ˈkɑːbaid/*n.* 碳化物，电石，碳化钙

25. compressive/kəmˈpresiv/*a.* 压缩的，有

26. epoxy resins 环氧树脂

27. （be）subjected to 曾受到，使受到

28. in some cases 有时，在有些情况下

Notes

[1] To properly select a tool material, there are several physical and mechanical properties you should understand to determine how the materials you select will affect the function and operation of the tool.

为了选择工具材料，你应当掌握材料的一些物理性能和力学性能，以便确定所选材料对工具的功能和操作会有何影响。

不定式 to properly select a tool material 放在句首用作目的状语；you select 为定语从句修饰前面的 the materials。

[2] A great effort is made to keep porosity, segregation, impurities, and nonmetallic inclusions to as low a level as possible.

（对于工具钢的冶炼，）要最大限度地降低钢中的气孔、偏析、杂质以及非金属夹杂物等的含量。

这里的 a great effort is made 是 make a great effort 的被动形式，意为"尽一切力量"。

Glossary of Terms

1. carbide tool 硬质合金刀具
2. alloy tool steel 合金工具钢
3. alloy cast iron 合金铸铁
4. carbon steel 碳素钢
5. carbon tool steel 碳素工具钢
6. cast iron 铸铁
7. cast steel 铸钢
8. die block steel 模具钢
9. die material 模具材料
10. free cutting steel 易切削钢
11. high alloy steel 高合金钢
12. high carbon steel 高碳钢
13. low alloy steel 低合金钢
14. low carbon steel 低碳钢
15. shock resistant tool steel 抗冲击工具钢
16. cold work tool（die）steel 冷作工具（模具）钢
17. hot work tool（die）steel 热作工具（模具）钢
18. nodular graphite iron 球墨铸铁
19. malleable cast iron 可锻铸铁
20. mottled cast iron 麻口铸铁
21. high-speed steel 高速钢
22. white cast iron 白口铸铁
23. compacted graphite cast iron 蠕墨铸铁
24. powder metallurgy（P/M）粉末冶金

Reading Materials

Carbon Steels

Carbon steels are used extensively in tool construction. Carbon steels are those steels which only contain iron and carbon, and small amounts of other alloying elements. Carbon steels are the most common and least expensive type of steel used for tools. The three principal types of carbon steels used for tooling are low carbon, medium carbon, and high carbon steels. Low carbon steel contains carbon between 0.05% and 0.3%. Medium carbon steel contains carbon between 0.3% and 0.7%. And high carbon steel contains carbon between 0.7% and 1.5%. As the carbon content is increased in carbon steel, the strength, toughness, and hardness are also increased when the metal is heat treated.

Low carbon steels are soft, tough steels that are easily machined and welded. Due to their low carbon content, these steels cannot be hardened except by case hardening. Low carbon steels are well suited for the following applications: tool bodies, handles, die shoes, and similar situations where strength and wear resistance are not required.

Medium carbon steels are used where greater strength and toughness is required. Since medium carbon steels have a higher carbon content, they can be heat treated to make parts such as studs, pins, axles, and nuts. Steels in this group are more expensive as well as more difficult to machine and weld than low carbon steels.

High carbon steels are the most hardenable type of carbon steel and are used frequently for parts where wear resistance is an important factor. Other applications where high carbon steels are well suited include drill bushings, locators, and wear pads. Since the carbon content of these steels is so high, parts made from high carbon steel are normally difficult to machine and weld.

Alloy Steels

Alloy steels are basically carbon steels with additional elements added to alter the characteris-tics and bring about a predictable change in the mechanical properties of the alloyed metal. Alloy steels are not normally used for most tools due to their increased cost, but some have been found favor for special applications. The alloying elements used most often in steels are manganese, nickel, molybdenum and chromium.

Another type of alloy steel frequently used for tooling applications is stainless steel. Stainless steel is a term used to describe high chromium and nickel-chromium steels. These steels are used for tools which must resist high temperatures and corrosive atmospheres. Some high chromium steels can be hardened by heat treatment and are used where resistance to wear, abrasion, and corrosion are required. Typical applications where a hardenable stainless steel is sometimes preferred are plastic injection molds. Here the high chromium content allows the steel to be highly polished and prevents deterioration of the cavity from heat and corrosion.

Die Steels

The die material is relevant directly to the die life as well as die cost. Recently, the commonly used die steel for cold blanking is as follows:

The carbon tool steel T8A, T10A, T12A are the cheapest and the most widely used. Its hardness after annealing is lower than that of the alloy steel. It has good machining property, and the process of forging, annealing and quenching is easy to be mastered. It is suitable to manufacture the working parts of the blanking die with small size and simple shape. But its hardenability and abrasion-resistant are bad, its quench distortion is large and its working life is short.

The low alloy tool steel, such as CrWMn, 9CrSi, 9Mn2V etc, can be quenched by oil, therefore it has a good hardenability and a small quench distortion. Comparing with T10A, 9Mn2V has a higher hardness and abrasion-resistance, and also has a good machining property.

High-carbon high-chrome die steel, such as Cr12, Cr12Mo, Cr12MoV etc, has a high strength, good hardenability and abrasion-resistance and small quench distortion. The carbon content of Cr12 is a bit higher. The distribution of carbide is nonuniform severely, which results in a decrease in the strength as well as hardness.

High-carbon medium-chrome die steel includes Cr6WV, Cr4W2MoV etc. Cr6WV is advantageous in small chrome content, better strength and impact toughness as compared with Cr12. Due to the small content of carbon and chrome, its abrasion-resistance and hardenability is not as good as Cr12, but its quench distortion is small. Its life expectancy is almost the same as Cr12. Cr4W2MoV is a new brand of die steel for cold forming to substitute Cr12. As compared with Cr12, it is characterized in finer size and more uniform distribution of eutectic carbide, therefore its hardenability, hardening-quenching capacity, mechanical property and abrasion-resistance are a bit higher. The alloy elements, such as W, Mo and V etc, improve the stability of the steel and make the die undergoing possible, chemical heat treatment possible.

Lesson 2   Heat Treating of Tool Steels

Text

The purpose of heat treatment is to control the properties of a metal or alloy through the *alteration* of the structure of the metal or alloy by heating it to definite temperatures and cooling at various rates. This combination of heating and controlled cooling determines not only the *nature* and *distribution* of the micro*constituents*, which in turn determine the properties, but also the *grain* size[1].

Heat treating should improve the alloy or metal for the service intended. Some of the various purposes of heat treating are as follows:

1. To remove *strains* after cold working.
2. To remove internal stresses such as those produced by drawing, bending, or welding.
3. To increase the hardness of the material.
4. To improve machinability.
5. To improve the cutting capabilities of tools.
6. To increase wear-resisting properties.
7. To soften the material, as in *annealing*.
8. To improve or change properties of a material, such as *corrosion* resistance, heat resistance, *magnetic* properties, or others as required.

Treatment of Ferrous Materials. Iron is the major constituent in the steels used in tooling, to which carbon is added in order that the steel may harden. Alloys are put into steel to enable it to develop properties not possessed by plain carbon steel, such as the ability to harden in oil or air, increased wear resistance, higher toughness, and greater safety in hardening.

Heat treatment of ferrous materials involves several important operations which are *customarily* referred to under various headings, such as normalizing, *spheroidizing*, stress relieving, annealing, hardening, tempering, and case hardening.

Normalizing involves heating the material to a temperature of about 100 ~ 200°F (55 ~ 100°C) above the critical range and cooling in still air. This is about 100°F (55°C) over the regular hardening temperature.

The purpose of normalizing is usually to refine grain structures that have been *coarsened* in forging. With most of the medium-carbon forging steels, alloyed and unalloyed, normalizing is highly recommended after forging and before machining to produce more *homogeneous* structures, and in most cases, improved machinability[2].

High-alloy air-hardened steels are never normalized, since to do so would cause them to harden and defeat the primary purpose.

Spheroidizing is a form of annealing which, in the process of heating and cooling steel, produces a rounded or *globular* form of carbide—the hard constituent in steel.

Tool steels are normally spheroidized to improve machinability. This is accomplished by heating to a temperature to 1380 ~ 1400°F (749 ~ 760°C) for carbon steels and higher for many alloy tool steels, holding at heat one to four hours, and cooling slowly in the furnace.

Stress Relieving. This is a method of relieving the internal stresses set up in steel during forming, cold working, and cooling after welding or machining. It is the simplest heat treatment and is accomplished merely by heating to 1200 ~ 1350°F (649 ~ 732°C) followed by air or furnace cooling.

Large dies are usually roughed out, then stress-relieved and finish-machined. This will minimize change of shape not only during machining but during subsequent heat treating as well. Welded sections will also have locked-in stresses owing to a combination of differential heating and cooling cycles as well as to changes in cross section. Such stresses will cause considerable movement in machining operations.

Annealing. The process of annealing consists of heating the steel to an *elevated* temperature for a definite period of time and, usually, cooling it slowly. Annealing is done to produce homogenization and to establish normal equilibrium conditions, with corresponding characteristic properties.

Tool steel is generally purchased in the annealed condition. Sometimes it is necessary to rework a tool that has been hardened, and the tool must then be annealed. For this type of anneal, the steel is heated slightly above its critical range and then cooled very slowly.

Hardening. This is the process of heating to a temperature above the critical range, and cooling rapidly enough through the critical range to *appreciably* harden the steel.

Tempering. This is the process of heating quenched and hardened steels and alloys to some temperature below the lower critical temperature to reduce internal stresses set up in hardening.

Case Hardening. The addition of carbon to the surface of steel parts and the subsequent hardening operations are important phases in heat treating. The process may involve the use of molten sodium *cyanide* mixtures, pack *carburizing* with activated solid material such as charcoal or coke, gas or oil carburizing, and dry cyaniding.

Questions

1. What process is used to remove the internal stresses created during a hardening operation?
2. What heat treating process makes the metallic carbides in a metal form into small rounded globules?
3. What are the main purposes of heat treating?
4. How many heat treating processes are involved in ferrous materials?

New Words and Expressions

1. alteration/ɔːltəˈreiʃən/*n.* 改变，变更
2. nature/ˈneitʃə/*n.* 本性，性质，自然界
3. distribution/distriˈbjuːʃən/*n.* 分配，分布
4. constituent/kənˈstitjuənt/*n.* 成分，分量，

要素；*a.* 组成的

5. grain/grein/*n.* 晶粒，粒度；*vt.* 使结晶，使成细粒；*vi.* 形成粒状

6. strain/strein/*vt. n.* 应变，张力，变形，弯曲

7. annealing/əˈniːliŋ/*n.* 退火，韧化，缓冷

8. corrosion/kəˈrəuʒən/*n.* 腐蚀，侵蚀，锈，铁锈

9. magnetic/mægˈnetik/*a.* 磁的，磁化的，有吸引力的

10. customarily/ˈkʌstəmərili/*ad.* 通常，习惯上

11. spheroidizing/sfiəˈrɔidaiziŋ/*n.* 球化处理

12. coarsen/ˈkɔːsn/*vt.* 使粗，粗化；*vi.* 变粗

13. homogeneous/hɔməˈdʒiːnjəs/*a.* 同种的，同性的，均匀的

14. globular/ˈglɔbjulə/*a.* 球状的，有小球的，世界范围的

15. elevate/ˈeliveit/*vt.* 抬起，举起，使升高

16. appreciable/əˈpriːʃiəbl/*a.* 可估计的，明显的，可观的

17. cyanide/ˈsaiənaid/*n.* 氰化物

18. carburize/ˈkɑːbjuraiz/*vt.* 渗碳

19. grain size   晶粒尺寸

20. cold working   冷加工

21. internal stress   内应力

22. corrosion resistance   耐腐蚀

23. heat resistance   耐热

24. magnetic property   磁性能

25. （be）referred to   把……归因于，参考，认为……由于

26. critical range   临界范围

Notes

[1] This combination of heating and controlled cooling determines not only the nature and distribution of the microconstituents, which in turn determine the properties, but also the grain size.

（在热处理过程中，）加热与冷却控制相结合的方法不仅决定了材料中微观组织的分布和性质（进而决定了该材料的性能），而且也决定了材料内部晶粒的大小。

句中第一个 and 连接的是 heating 和 controlled cooling，第二个 and 连接的是 nature 和 distribution，而 which in turn determine the properties 为非限定性定语从句，修饰 the nature and distribution。

[2] With most of the medium-carbon forging steels, alloyed and unalloyed, normalizing is highly recommended after forging and before machining to produce more homogeneous structures, and in most cases, improved machinability.

对于大多数中碳锻钢来说，不管是否合金化，在锻造后和机械加工前通常推荐采用正火处理工艺，这样有利于形成更均匀的组织，并且在大多数情况下可改善材料的切削加工性能。

句首的介词 with 短语意为"对……来说"；in most cases 可译为"在大多数情况下"。

Glossary of Terms

1. hardenability   淬透性

2. hardenability curve   淬透性曲线

3. hardening capacity   淬硬性（硬化能力）

4. case hardening   表面硬化，渗碳

5. hardness profile 硬度分布（硬度梯度）

6. heat treatment procedure 热处理规范

7. heat treatment installation 热处理设备

8. heat treatment furnace 热处理炉

9. heat treatment cycle 热处理工艺周期

10. heat time 加热时间

11. heat system 加热系统

12. heating up time 升温时间

13. heating curve 加热曲线

14. high temperature carburizing 高温渗碳

15. high temperature tempering 高温回火

16. isothermal transformation 等温转变

17. isothermal annealing 等温退火

18. interrupted ageing treatment 分级时效处理

19. local heat treatment 局部热处理

20. overheated structure 过热组织

21. pack carburizing 固体渗碳

22. oxynitrocarburizing 氧氮碳共渗

23. partial annealing 不完全退火

24. spheroidized structure 球化组织

25. recrystallization temperature 再结晶温度

Reading Materials

Heat Treatment of Die Steels

Although alloy steels contain elements such as chromium, molybdenum and vanadium, two constituents are essential for heat treatment: iron, termed ferrite in metallography, and carbon, which combines with iron to form cementite, the hard intermetallic compound Fe_3C. These two constituents form a eutectoid structure known as pearlite when the steel is cooled slowly enough to reach equilibrium, but by rapid cooling the steel is hardened. When such a quenched steel is tempered, structures with mechanical properties intermediate between those of the slowly cooled and the quenched conditions are formed.

In recent years there has been a greater understanding of the complex structural changes taking place during heat treatment, with the help of phase transformation diagrams. Use of these diagrams can lead to better control of the heat treatment cycle which in turn will ensure that optimum properties and maximum die life are achieved.

Surface Treatments for Steels

During the past 20 years, several processes have been introduced to obtain enhanced surface hardness of steels. Some of them have developed from case carburizing and nitriding, to obtain shorter processing times with better environmental control and improved properties. Various salt bath processes have been used and now a wide range of new methods is available.

In the die casting industry surface treatments are applied to steels to improve the properties of nozzles, ejector pins, cores and shot sleeves, to provide maximum resistance to erosion, pitting and soldering. Treatment of die cavities has received only limited acclaim, because the complex thermal patterns produced on large die components lead to stresses which are sufficiently high to break through the thin surface treated layers, leading to premature failure. Experience in drop forging has also indicated that surface treatments of their dies have not been particularly successful.

Thermochemical treatments are applied to die casting die components; the surface chemistry of the steel is modified by the introduction of nitrogen, carbon and sometimes other elements; the processes are of the main types listed below.

1. Nitriding.
2. Nitrocarburizing such as Tufftride, Sulfinuz and Suraulf.
3. Metallizing such as boronizing and the Toyota diffusion process.
4. Carburizing and carbonitriding.

Lesson 3   Third–Angle *Projection*

Text

The Six *Views*. Any object can be viewed from six *mutually perpendicular* directions, as shown in Figure 3-1a. These six views may be *drawn* if necessary, as shown in Figure 3-1b. The six views are always *arranged* as shown, which is the American National Standard arrangement. The *top*, *front*, and bottom views *align* vertically, while the *rear*, left-side, front, and right-side views align horizontally. To draw a view out of place is a serious error and is generally regarded as one of the worst possible mistakes in drawing[1].

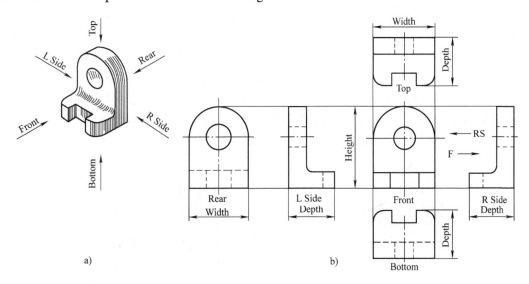

a)   b)

Figure 3-1   The six views

Note that height is shown in the rear, left-side, front, and right-side views; width is shown in the rear, top, front, and bottom views; and depth is shown in the four views that surround the front view—namely, the left-side, top, right-side, and bottom views. Each view shows two of the principal dimensions. Note also that in the four views that surround the front view, the front of the object faces toward the front view.

Adjacent views are *reciprocal*. If the front view in Figure 3-1 is imagined to be the object itself, the right-side view is obtained by looking toward the right side of the front view, as shown by the *arrow* RS. Likewise, if the right-side view is imagined to be the object, the front view is obtained by looking toward the left side of the right-side view, as shown by the arrow F, the same relation exists between any two adjacent views.

Necessary Views. A drawing for use in production should contain only those views needed for a clear and complete shape *description* of the object. These minimum required views are referred to as the necessary views. In selecting views, the drafter should choose those that best show essential contours or shapes and have the least number of hidden lines.

As shown in Figure 3-1, three *distinctive* features of this object need to be shown on the drawing: ①rounded top and hole, seen from the front; ②rectangular notch and rounded corners, seen from the top; ③right angle with filleted corner, seen from the side.

The three principal dimensions of an object are width, height, and depth. In technical drawing, these fixed terms are used for dimensions taken in these directions, regardless of the shape of the object[2]. The terms "length" and "thickness" are not used because they cannot be applied in all cases. The top, front, and right-side views, arranged closer together, are shown in Figure 3-1. These are called the three regular views because they are the views most frequently used.

Alignment of Views. Errors in arranging the views are so commonly made by students that it is necessary to repeat this: The views must be drawn in accordance with the American National Standard arrangement shown in Figure 3-1. Figure 3-2a shows an offset guide that requires three views. These three views, correctly arranged, are shown in Figure 3-2b. The top view must be directly to the right of the front view—not out of alignment, as in Figure 3-2c. Also, never draw the views in reversed positions, with the bottom over the front or the right-side to the left of the front view (Figure 3-2d), even through the views do line up with the front view.

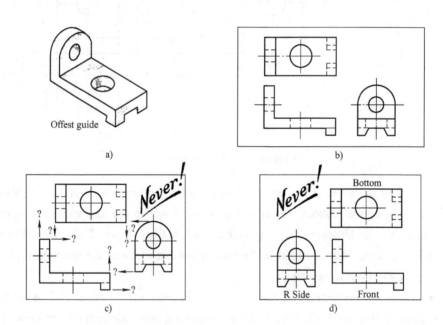

Figure 3-2   Position of views

Questions

1. What is the three-angle projection?
2. What are the differences between three-angle projection and first-angle projection?

3. List the six principal views of an object.

4. In a drawing that shows the top, front, and right-side view, which two views show depth? Which view shows depth vertically? Which view shows depth horizontally?

5. What are the three principal dimensions of an object?

New Words and Expressions

1. projection/prə'dʒekʃən/*n*. 投影，发射

2. view/vjuː/*n*. 视图；*vt*. 观察

3. mutually/'mjuːtjuəli/*ad*. 相互地，共同地

4. perpendicular/'pəːpən'dikjulə/*a*. 垂直的，正交的；*n*. 垂直，正交

5. draw/drɔː/*vt*. 拉，拖，绘制，描写；*vi*. 制图

6. arrange/ə'reindʒ/*vt*. 整理，排列，安排；*vi*. 安排，准备

7. top/tɔp/*n*. 顶端，顶部，上部；*a*. 最高的，顶上的

8. front/frʌnt/*n*. 正面，前面，前方；*a*. 正面的，前面的；*vt*. & *vi*. 面向，面对

9. align/ə'lain/*vt*. & *vi*. 使成一直线，排列成一行

10. rear/riə/*n*. 后部，后面；*a*. 后面的，后部的

11. adjacent/ə'dʒeisənt/*a*. 接近的，毗邻的

12. reciprocal/ri'siprəkəl/*a*. 相互的，相应的；*n*. 倒数

13. arrow/'ærəu/*n*. 箭，指针，箭号

14. description/dis'kripʃən/*n*. 叙述，图说，绘制

15. distinctive/dis'tiŋktiv/*a*. 有区别的，特殊的

16. American National Standard arrangement 美国国家标准排列

17. left-side view 左侧（视）图

18. right-side view 右侧（视）图

19. hidden line (dotted line, dashed line) 隐藏线，虚线

20. be generally regarded as 一般地被看作……，被认为……

21. look toward 面朝，期待；为了……做好准备

22. be referred to as 称为，被认为是

23. three regular views 三个常规视图，三视图

24. in accordance with 按照，依据，与……一致

25. line up with 排成一行

Notes

[1] To draw a view out of place is a serious error and is generally regarded as one of the worst possible mistakes in drawing.

将视图绘制在不适当的位置是一个严重的错误，而且常常被认为是绘图过程中可能出现的最为严重的错误之一。

To draw a view out of place 为不定式短语，在句中作主语；句中 out of place 可译为"不合适，不在适当的位置"；is generally regarded as 可译为"常常被认为……"。

[2] In technical drawing, these fixed terms are used for dimensions taken in these directions, regardless of the shape of the object.

不论物体的形状如何，在技术制图中，这些固定术语常常用来表述这些方向的尺寸。In technical drawing 在句中作状语，可译为"在技术制图中"；regardless of 作"不管""不顾""不论……如何"解。

Glossary of Terms

1. three-angle projection　三角投影
2. third-angle projection　第三角投影
3. first-angle projection　第一角投影
4. drafting　制图
5. mechanical drawing　机械制图
6. standard drawing　标准图
7. standard components（parts）　标准件
8. the sheet, drawing sheet, drawing paper　图样
9. drawer, draftsman　绘图员
10. working drawing　工作图，生产图
11. detail drawing, part drawing　零件图
12. sketch（layout, outline）　草图
13. assembly drawing　装配图
14. design drawing　设计图
15. blueprint　蓝图
16. engineering drawing, structure drawing　工程图，结构图
17. machine parts, mechanical device, components　零件，部件
18. title block　标题栏
19. sectional view　断面图
20. orthographic projection　正投影
21. the top view　俯视投影，俯视图
22. the front view　主视投影（主视图，正面图）
23. the side view　侧投影
24. the bottom view　仰视图
25. front view　主视图，前视图
26. rear（back）view　后视图
27. side（end）view　侧（端）视图
28. three-view drawing　三视图
29. profile, section（full ~ , half ~ , offset ~ , broken-out ~ , rotating ~ , inclined ~ , compound ~ ）　剖面（全剖，半剖，阶梯剖，局部剖，旋转剖，斜剖，复合剖）
30. technical requirements　技术要求
31. a detail list of components, the part list of an assembly drawing　零件明细表
32. scale, proportional scale　比例
33. dimensional line　尺寸线
34. descriptive geometry　画法几何
35. dimensioning, to indicate the sizes on the sheet, size marking　尺寸标注

Reading Materials

First-Angle Projection

If the vertical and horizontal planes of projection are considered indefinite in extent and intersecting at 90° with each other, the four dihedral angles produced are the first, second, third, and fourth angles (Figure 3-3a).

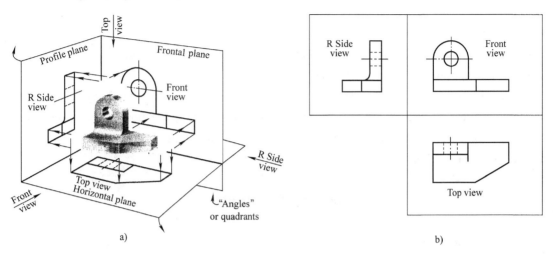

Figure 3-3　First-angle projection

If the object is placed above the horizontal plane and in front of the vertical plane, the object is in the first angle. In this case, the observer always looks through the object and to the planes of projection. Thus, the right-side view is still obtained by looking toward the right side of the object, the front by looking toward the front, and the top by looking down toward the top; but the views are projected from the object onto a plane in each case. When the planes are unfolded (Figure 3-3b), the right-side view falls at the left of the front view, and the top view falls below the front view, as shown. A comparison between first-angle orthographic projection and third-angle orthographic projection is shown in Figure 3-4. The front, top, and right-side views shown in Figure 3-3b for first-angle projection are repeated in Figure 3-4a. Ultimately, the only difference between third-angle and first-angle projection is the arrangement of the views. Still, confusion and

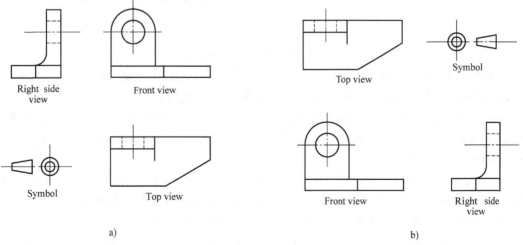

Figure 3-4　First-angle projection compared to third-angle projection

a) First-angle projection　b) Third-angle projection

possibly manufacturing errors may result when the user reading a first-angle drawing thinks it is a third-angle drawing, or vice versa. To avoid misunderstanding, international projection symbols, shown in Figure 3-4, have been developed to distinguish between first-angle and third-angle projections on drawings. On drawings where the possibility of confusion is anticipated, these symbols may appear in or near the title box.

In the United States and Canada (and, to some extent, in England), third-angle projection is standard, while in most of the rest of the world, first-angle projection is used. First-angle projection was originally used all over the world, including the United States, but it was abandoned around 1890.

Lines in Sectioning

A correct front view and sectional view are shown in Figure 3-5a and Figure 3-5b. In general, all visible edges and contours behind the cutting plane should be shown; otherwise a section will appear to be made up of disconnected and unrelated parts, as shown in Figure 3-5c. Occasionally, however, visible lines behind the cutting plane are not necessary for clarity and should be omitted.

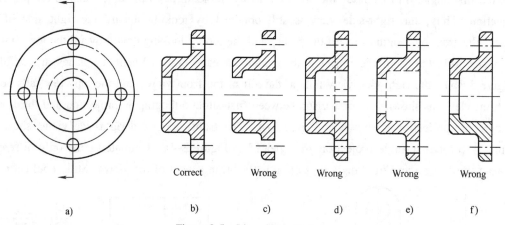

Figure 3-5   Lines in sectioning

Sections are used primarily to replace hidden-line representation, and, as a rule, hidden lines should be omitted in sectional views. As shown in Figure 3-5d, the hidden lines do not clarify the drawing; they tend to confuse, and they take unnecessary time to draw. Sometimes hidden lines are necessary for clarity and should be used in such cases, especially if their use will make it possible to omit a view.

A section-lined area is always completely bounded by a visible outline, never by a hidden line, as in Figure 3-5e, since in every case the cut surfaces and their boundary lines will be visible. Also, a visible line can never cut across a section-lined area.

In a sectional view of an object, alone or in assembly, the section lines in all sectioned areas must be parallel, not as shown in Figure 3-5f. The use of section lining in opposite directions is an indication of different parts, as when two or more parts are adjacent in an assembly drawing.

Lesson 4　Tolerance

Text

Interchangeable manufacturing allows parts made in widely separated localities to be brought together for *assembly*. That the parts all fit together properly is an essential element of mass production. Without interchangeable manufacturing, modern industry could not exist, and without effective size control by the engineer, interchangeable manufacturing could not be achieved[1].

However, it is impossible to make anything to exact size. Parts can be made to very close dimensions, even to a few millionths of an inch or thousandths of a *millimeter*, but such *accuracy* is extremely expensive.

Fortunately, exact sizes are not needed. The need is for varying degrees of accuracy according to functional requirements. A manufacturer of children's *tricycles* would soon go out of business if the parts were made with jet-engine accuracy—no one would be willing to pay the price[2]. So what is wanted is a means of specifying dimensions with whatever degree of accuracy is required. The answer to the problem is the specification of a *tolerance* on each dimension.

Tolerance is the total amount that a specific dimension is permitted to vary; it is the difference between the maximum and the minimum *limits* for the dimension. It can be specified in any of the two forms: *unilateral* or *bilateral*. In unilateral tolerance, the variation of the size will be *wholly* on the side. For example, $30_{-0.02}^{\ 0}$ is a unilateral tolerance. Here the nominal dimension 30 is allowed to vary between 30mm and 29.98mm. In bilateral tolerance, the variation will be to both the sides. For example, 30.00 ± 0.01 or $30_{-0.10}^{+0.05}$. In bilateral tolerance, the variation of the limits can be uniform as shown in the former case. The dimension varies from 30.01mm to 29.99mm. *Alternatively* the allowed *deviation* can be different as shown in the second case. Here the dimension varies from 30.05mm to 29.90mm.

In engineering when a product is designed it consists of a number of parts and these parts mate with each other in some form. In the assembly it is important to consider the type of mating or fit between two parts which will actually define the way the parts are to behave during the working of the assembly.

Take for example a shaft and hole, which will have to fit together. In the simplest case if the dimension of the shaft is lower than the dimension of the hole, then there will be *clearance*. Such a fit is termed clearance fit. Alternatively, if the dimension of the shaft is more than that of the hole, then it is termed *interference* fit. However, depending upon the possibilities of dimensions, at times there will be clearance and other times there will be interference. Such a fit is termed as *transition* fit. These are illustrated in Figure 4-1.

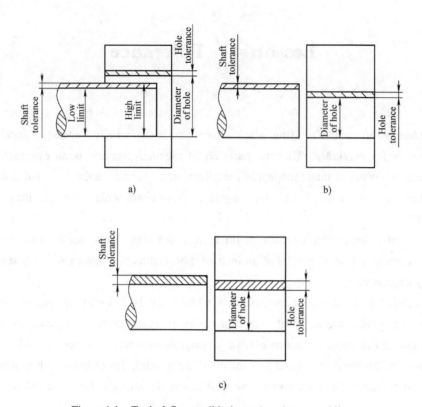

Figure 4-1 Typical fits possible in engineering assemblies
a）Clearance fit　b）Interference fit　c）Transition fit

Questions

1. Why is it impossible to make anything to exact size?
2. What is tolerance?
3. What is the difference between unilateral tolerance and bilateral tolerance?
4. Explain the concepts of clearance, interference and transition fits.

New Words and Expressions

1. interchangeable/ˌintəˈtʃeindʒəbl/ *a*. 可交换的，可互换的　interchangeability *n*. 互换性
2. assembly/əˈsembli/ *n*. 集合，集会；装配，部件，组件；装配图，总图
3. millimeter/ˈmiliˌmiːtə/ *n*. 毫米
4. accuracy /ˈækjurəsi/ *n*. 准确（度），精确（度）
5. fortunately/ˈfɔːtʃənətli/ *ad*. 幸运地，侥幸地
6. tricycle /ˈtraisikl/ *n*. 三轮车
7. tolerance/ˈtɔlərəns/ *n*. 公差，容差
8. limit/ˈlimit/ *n*. 界限，限度，范围；极限，公差；*vt*. 限制，限定，减小
9. unilateral/ˌjuːniˈlætərəl/ *a*. 单边的，一方的，单向的
10. bilateral/baiˈlætərəl/ *a*. 双边的，两边的
11. wholly/ˈhəuli/ *ad*. 完全地，实足，统统
12. alternatively/ɔːlˈtəːnətivli/ *ad*. 两者挑一地，交替地；可能地，选择地

13. deviation /diːviˈeiʃən/ *n.* 偏差，偏离，偏向

14. clearance /ˈkliərəns/ *n.* 间隙，空隙，公隙

15. interference /intəˈfiərəns/ *n.* 过盈，干涉，冲突，抵触

16. transition /trænˈsiʃən/ *n.* 过渡；转变，转移，

17. tolerance on fit 配合公差

18. tolerance zone 公差带

19. mass production 成批生产，大批生产，批量生产

20. as a means of 作为……的工具（或方法，手段）

21. be termed as 被叫作，被称作

22. in the former case 在前一种情形下

Notes

[1] Without interchangeable manufacturing, modern industry could not exist, and without effective size control by the engineer, interchangeable manufacturing could not be achieved.

没有可互换性制造，现代工业就不可能存在；没有工程师对零件尺寸的有效控制，可互换性制造也就不可能实现。

该句中由介词 without 引出两个假设条件从句，分别在句中作状语，without 作"没有"解。

[2] A manufacturer of children's tricycles would soon go out of business if the parts were made with jet-engine accuracy—no one would be willing to pay the price.

如果童车制造商将童车制造成与喷气式发动机一样的精度，制造商将会很快退出童车市场，因为没有人情愿支付昂贵的价格购买童车。

本句中的 if 引导的是与现在事实相反的虚拟从句。

Glossary of Terms

1. unilateral tolerance 单边间隙

2. bilateral tolerance 双边间隙

3. clearance fit 间隙配合

4. interference fit 过盈（静）配合

5. transition fit 过渡配合

6. hole-basis (basic-hole) system 基孔制

7. shaft-basis (basic-shaft) system 基轴制

8. basic size 基本尺寸

9. actual size 实际尺寸

10. limit of size 极限尺寸

11. upper (lower) derivation 上（下）偏差

12. error 误差

13. standard tolerance 标准公差

14. tolerance grade 公差等级

15. nominal error 名义误差

16. geometry tolerance 形位公差（几何公差）

17. working (finishing) allowance 加工余量

18. straightness, flatness, circularity, cylinderity, parallelism, perpendicularity 直线度，平面度，圆度，圆柱度，平行度，垂直度

19. angularity, concentricity, symmetry, roughness 倾斜度，同轴度，对称度，表面粗糙度

20. total runout (runout) 全跳动（圆跳动）

21. through hole (blind hole) 通孔（不通孔）

22. chamfer, fillet angle　倒角，圆角

23. pin hole　销孔

24. datum（ ～line, ～plane）　基准（基准线，基准面）

25. straight line（arc, curve）　直线（圆弧，曲线）

26. horizontal line（incline line, vertical line）　水平线（斜线，垂直线）

27. continuous thick line（full line, visible line）　粗实线（实线，可见轮廓线）

28. continuous thin line　细实线

29. curve of intersection between two bodies　相贯线

30. standard　标准

31. GB　中华人民共和国强制性国家标准（简称国标）

32. GB/T　中华人民共和国推荐性国家标准

33. International Standardization Organization（ISO）　国际标准化组织

34. British Standard（BS）　英国标准

35. British Association of Standard（BA）　英国标准协会

36. American Standard（AS）　美国标准

37. American Standard Association（ASA）　美国标准协会

38. Japanese Standard（JS）　日本标准

39. Japanese Industrial Standard（JIS）　日本工业标准

40. Deutsches Institut für Normung（DIN）　德国标准化学会（德国工业标准）

Reading Materials

Hole-Basis and Shaft-Basis System

For obtaining the required fit, the organization can choose any one of the following two possible systems.

Hole-basis system. In this system the nominal size and the limits on the hole are maintained constantly and the shaft limits are varied to obtain the requisite fit. For example：

Let the hole size be $20.00^{+0.03}_{0}$.

Shaft of $20.00^{+0.08}_{+0.04}$ gives the interference fit.

Shaft of $20.00^{+0.04}_{0}$ gives the transition fit.

Shaft of $20.00^{-0.02}_{-0.04}$ gives the clearance fit.

Shaft-basis system. This is the reverse of hole-basis system. In this system the shaft size and limits are maintained constant while the limits of hole vary to obtain any fit.

Though there is not much to choose between the two systems, the hole-basis system is mostly used because standard tools such as reamers, drills, broaches and other standard tools are often used to produce holes, and standard plug gages are used to check the actual sizes. On the other hand, shafting can easily be machined to any size desired.

Preferred Fits

The symbols for either the hole-basis or shaft-basis preferred fits (clearance, transition, and interference) are given in table. Fits should be selected from Table 4-1 for mating parts where possible.

Table 4-1 Preferred fits

Fits	ISO symbol		Description
	Hole-basis	Shaft-basis	
Clearance fits	H11/c11	C11/h11	Loose-running fit for wide commercial tolerances or allowances on external members
	H9/d9	D9/h9	Free-running fit not for use where accuracy is essential, but good for large temperature variation, high running speeds, or heavy journal pressures
	H8/f7	F8/h7	Close-running fit for running on accurate machines and for accurate location at moderated speeds and journal pressures
Transition fits	H7/g6	G7/h6	Sliding fit not intended to run freely, but to move and turn freely and locate accurately
	H7/h6	H7/h6	Locational clearance fit provides snug fit for locating stationary parts; but can be freely assembled and disassembled
	H7/k6	K7/h6	Locational transition fit for accurate location, a compromise between clearance and interference
	H7/n6	N7/h6	Locational transition fit for more accurate location where greater interference is permissible
Interference fits	H7/p6	P7/h6	Locational interference fit for parts requiring rigidity and alignment with prime accuracy of location but without special bore pressure requirements
	H7/s6	S7/h6	Medium drive fit for ordinary steel parts or shrink fits on light sections, the tightest fit usable with cast iron
	H7/u6	U7/h6	Force fit suitable for parts which can be highly stressed or for shrink fits where the heavy pressing forces required are impractical

Although second and third choice basic size diameters are possible, they must be calculated from tables not included in this text. For the generally preferred hole-basis system, note that the ISO symbols range from H11/c11 (loose running) to H7/u6 (force fit). For the shaft-basis system, the preferred symbols range from C11/h11 (loose fit) to U7/h6 (force fit).

Unit Two

Lesson 5   Cutting Tool Design

Text

Physics of metal-cutting provide the theoretical *framework* by which we must examine all other elements of cutting tool design. We have workpiece materials from a very soft, *buttery* consistency to very hard and shear resistant. Each of the workpiece materials must be *handled* by itself; the amount of broad information that is applicable to each workpiece material is reduced as the *distinctions* between workpiece characteristics increase. Not only is there a vast *diversity* of workpiece materials, but there is also a variety of shapes of tools and tool compositions.

The tool designer must *match* the many variables to provide the best possible cutting *geometry*. There was a day when *trial* and error was normal for this decision, but today, with the ever-increasing variety of tools, trial and error is far too expensive.

The designer must develop expertise in applying data and making *comparisons* on the basis of the experience of others. For example: tool manufacturers and material salesmen will have figures when their companies have developed. The figures are meant to be guidelines; however, a careful examination of the *literature* available will provide an excellent place from which to start, and be much cheaper than trial and error.

Material removal by machining involves *interaction* of five elements: the cutting tool, the toolholding and guiding device, the workholder, the workpiece, and the machine. The cutting tool may have a single cutting edge or may have many cutting edges. It may be designed for linear or rotary motion. The geometry of the cutting tool depends upon its intended function. The toolholding device may or may not be used for guiding or locating. Toolholder selection is *governed* by tool design and intended function.

The physical composition of the workpiece greatly influences the selection of the machining method, the tool composition and geometry, and the rate of material removal[1]. The intended shape of the workpiece influences the selection of the machining method and the choice of linear or rotary tool travel. The composition and geometry of the workpiece to a great extent determines the workholder requirements. Workholder selection also depends upon forces produced by the tool on the workpiece. Tool guidance may be incorporated into the workholding function.

Successful design of tools for the material removal processes requires, above all, a complete understanding of cutting tool function and geometry. This knowledge will enable the designer to specify the correct tool for a given task. The tool, in turn, will govern the selection of toolholding and guidance methods. Tool forces govern selection of the workholding device. Although the process involves interaction of the five elements, everything begins with and is based on what happens at the point of contact between the workpiece and the cutting tool.

The primary method of imparting form and dimension to a workpiece is the removal of material by the use of edged cutting tools. An oversize mass is literally carved to its intended shape. The removal of material from a workpiece is termed generation of form by machining, or simply machining.

Form and dimension may also be achieved by a number of alternative processes such as hot or cold extrusion, sand casting, die casting, and precision casting. Sheet metal can be formed or drawn by the application of pressure. In addition to machining, metal removal can be accomplished by chemical or electrical methods. A great variety of workpieces may be produced without resorting to a machining operation. Economic considerations, however, usually dictate form generation by machining, either as the complete process or in conjunction with another process.

Cutting tools are designed with sharp edges to minimize rubbing contact between the tool and workpiece. Variations in the shape of the cutting tool influence tool life, surface finish of the workpiece, and the amount of force required to shear a chip from the parent metal. The various angles on a tool compose what is often termed the tool geometry. The tool *signature* or *nomenclature* is a sequence of *alpha* and *numeric* characters representing the various angles, *significant* dimensions, special features, and the size of the *nose* radius[2]. This method of *identification* has been standardized by the American National Standards Institute for carbide and for high speed steel, and is illustrated in Figure 5-1, together with the elements that make up the tool signature.

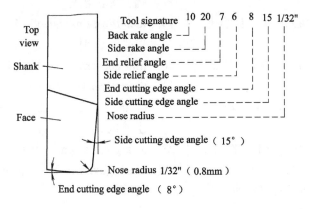

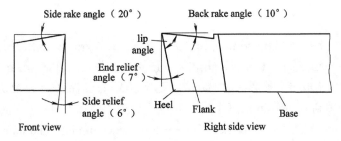

Figure 5-1　A straight-shank, right-cut, single-point tool, illustrating the elements of the tool signature as designated by the ANSI. Positive rake angles are shown

Questions

1. What five elements interact in the machining process?
2. What factor will greatly influence the cutting tool material and geometry?
3. What factors are influenced by the shape of a cutting tool?

New Words and Expressions

1. framework/ˈfreimwəːk/n.　骨架，构架，框架，结构，机构，组织
2. buttery/ˈbʌtəri/a.　黄油状的，涂有黄油的
3. handle/ˈhændl/n.　柄，把手；vt.　管理，处理，触，摸，应付；vi.　易于操纵
4. distinction/disˈtiŋkʃən/n.　差别，区别，特征，特性
5. diversity/daiˈvəːsiti/n.　参差，不同，多样性，发散
6. match/mætʃ/n.　比赛，匹配；vt.　配得上，搭配，和……相称；vi.　相适合
7. geometry/dʒiˈɔmitri/n.　几何学，几何形状，几何图
8. trial/ˈtraiəl/n.　试验，尝试　a.　试验性的，试制的
9. comparison/kəmˈpærisn/n.　比较，对照，比喻
10. literature/ˈlitərəˌtʃə/n.　文学，文献，著作，文学作品
11. interaction/intərˈækʃən/n.　相互作用，相互影响，相互制约
12. govern/ˈgʌvən/vt.　统治，管理，决定，控制；vi.　统治，管理
13. signature/ˈsignətʃə/n.　签名，署名，用法说明
14. nomenclature/nəuˈmenklətʃə/n.　名称，术语，命名法，专门用语
15. alpha/ˈælfə/n.　希腊字母"A，α"，最初，开始
16. numeric/njuːˈmerik/a.　数字的，数值的；n.　数，分数
17. significant/sigˈnifikənt/a.　有意义的，重要的，重大的，有效的
18. nose/nəuz/n.　头部，机头，管口，喷嘴，突出部分
19. identification/aiˌdentifiˈkeiʃən/n.　辨别，鉴别，识别
20. trial and error　反复试验
21. theoretical framework　理论框架
22. on the basis of　在……基础上，根据，主要成分
23. to a great extent　很大程度上
24. (be) governed by　取决于，以……为转移

Notes

[1] The physical composition of the workpiece greatly influences the selection of the machining method, the tool composition and geometry, and the rate of material removal.
实际工件的组成极大地影响加工方法、刀具组成与几何形状，以及切削速度的选择。
此句两处用到 composition 这个词，其意思是指工件和刀具的"组成成分"。

[2] The tool signature or nomenclature is a sequence of alpha and numeric characters representing the various angles, significant dimensions, special features, and the size of the nose radius.

刀具的命名是以希腊字母"α"和一些数字构成的一个有序排列，其中的数字代表着一些刀具的角度、重要尺寸、特殊性能以及刀尖半径的大小。

句中由 representing 引导的分词短语作后置定语，用来修饰 a sequence of alpha and numeric characters。

Glossary of Terms

1. cutting part　切削部分
2. cutting power　切削功率
3. cutting tooling　切削加工工艺装备
4. cutting force　切削力
5. cutting edge　切削刃
6. cutting speed　切削速度
7. depth of cut　背吃刀量
8. tool angle　刀具角度
9. tool back rake　背前角
10. tool back clearance　背后角
11. tool backlash movement（tool retracting）退刀
12. tool back wedge angle　背楔角
13. tool base clearance　基后角
14. tool center point　刀具中心
15. tool cutting edge angle　主偏角
16. tool cutting edge plane　切削平面
17. tool element（dimensions）　刀具要素（尺寸）
18. tool function　刀具功能
19. tool grinding machine　工具磨床
20. tool holder　刀夹
21. tool post（tool rest）　刀架
22. tool geometrical rake　几何前角
23. tool normal clearance（rake）　法后角（法前角）
24. tool offset　刀具偏置
25. tool orthogonal clearance（rake, wedge）后角（前角，楔角）
26. form tool　成形刀具

Reading Materials

Chip Formation

The majority of metal-cutting operations involve the separation of small segments or chips from the workpiece to achieve the required shape and size of manufactured parts. Chip formation involves three basic requirements：（1）there must be a cutting tool that is harder and more wear-resistant than the workpiece material；（2）there must be interference between the tool and the workpiece as designated by the feed and depth of cut；（3）there must be a relative motion or cutting velocity between the tool and the workpiece with sufficient force to overcome the resistance of the workpiece material. As long as these three conditions exist, the portion of the material being machined that interferes with free passage of the tool will be displaced to create a chip.

Many possibilities and combinations exist that may fulfill such requirements. Variations in tool material and tool geometry, feed and depth of cut, cutting velocity, and workpiece material have an effect not only upon the formation of the chip, but also upon cutting force, cutting horsepower, cutting temperatures, tool wear and tool life, dimensional stability, and the quality of the newly created surface. The interrelationship and the interdependence among these "manipulating factors" constitute the basis for the study of machinability—a study which has been popularly defined as the response of a material to machining.

Figure 5-2 illustrates the necessary relationship between cutting tool and the workpiece for chip formation in several common machining processes. Although it is apparent that different general shapes and sizes of chips may be produced by each of the basic processes, all chips regardless of process are usually classified according to their general behavior during formation.

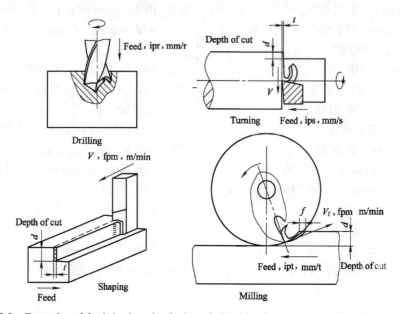

Figure 5-2   Examples of feed depth and velocity relationships for several chip-formation processes

Tool Life

The types and mechanisms of tool failure have been previously described. It was shown that excessive cutting speeds cause a rapid failure of the cutting edge; thus, the tool can be declared to have a short life. Other criteria are sometimes used to evaluate tool life:

1. Change of quality of the machined surface.

2. Change in the magnitude of the cutting force resulting in changes in machine and work-piece deflections causing workpiece dimensions to change.

3. Change in the cutting temperature.

4. Costs, including labor costs, tool costs, tool changing time (cost), etc.

The selection of the correct cutting speed has an important bearing on the economics of all metal cutting operations. Fortunately, the correct cutting speed can be estimated with reasonable accuracy from tool-life graphs or from the Taylor Tool Life Relationship, provided that necessary data are obtainable.

Lesson 6   Workholding Principles

Text

The term *workholder* includes all devices that hold, *grip*, or *chuck* a workpiece to perform a manufacturing operation. The holding force may be applied mechanically, electrically, hydraulically, or *pneumatically*. This section considers workholders used in material-removing operations. Workholding is one of the most important elements of machining processes.

Figure 6-1 illustrates almost all the basic elements that are present in a material-removing operation intended to shape a workpiece. The right hand is the toolholder, the left hand is the workholder, the knife is the cutting tool, and the piece of wood is the workpiece. Both hands combine their motion to shape the piece of wood by removing material in the form of chips. The body of the person whose hands are shown may be considered a machine that imparts power, motion, position, and control to the elements shown. Except for the element of force multiplication, these basic elements may be found in all of the forms of manufacturing setups where toolholders and workholders are used.

Figure 6-2 shows a pair of *pliers* or *tongs* used to hold a rod on which a point has to be ground or filed. This simple workholder illustrates the element of force multiplication by a lever action, and also shows *serrations* on the parts contacting the rod to increase resistance against *slippage*.

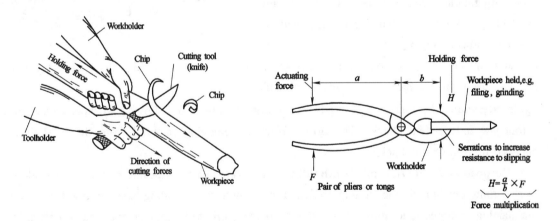

Figure 6-1   Principles of workholders   Figure 6-2   Multiplication of holding force

$$H = \frac{a}{b} \times F$$
Force multiplication

Figure 6-3 shows a widely used workholder, the screw-operated *vise*. The screw pushes the movable *jaw* and multiplies the applied force. The vise remains locked by the self-locking characteristic of the screw, provides means of *attachment* to a machine, and permits precise placement of the work[1].

A vise with a number of refinements often used in workholders is **depicted** in Figure 6-4. The main holding force is supplied by hydraulic power, the screw being used only to bring the jaws in contact with a workpiece. The jaws may be replaceable inserts profiled to locate and fit a specific workpiece as shown. Other, more complicated jaw forms are used to match more complicated workpieces.

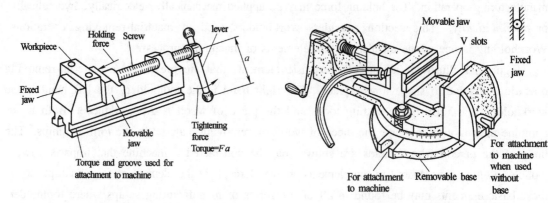

Figure 6-3   Elementary workholder (vise)   Figure 6-4   Vise with hydraulic clamping

Another large group of workholders are the chucks. They are attached to a variety of machine tools and are used to hold a workpiece during turning, boring, drilling, grinding, and other rotary operations. Many types of chucks are available. Some are tightened manually with a **wrench**, others are power operated by air or hydraulic means or by electric motors. On some chucks, each jaw is individually advanced and tightened, while others have all jaws advance in **unison**. Figure 6-5 shows a workpiece clamped in a four-jaw independent chuck. The drill, which is removing material from the workpiece is clamped in a universal chuck.

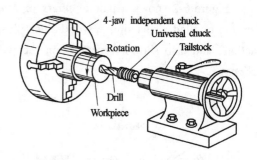

Figure 6-5   Holding (chucking) a round workpiece

Purpose and Function of Workholders. A workholder must position or locate a workpiece in a definite relation to the cutting tool and must withstand holding and cutting forces while maintaining that precise location. A workholder is made up of several elements, each performing a certain function. The locating elements position the workpiece; the structure, or tool body, withstands the forces; brackets attach the workholder to the machine; and clamps, screws, and jaws apply holding forces. Elements may have manual or power activation. All functions must be performed with the required **firmness** of holding, accuracy of positioning, and with a high degree of safety for the operator and the equipment.

The design or selection of a workholder is governed by many factors, the first being the physical characteristics of the workpiece. The workholder must be strong enough to support the workpiece without deflection. The workholder material must be carefully selected with the workpiece in mind so that neither will be damaged by *abrupt* contact, e. g., damage to a soft copper workpiece by hard steel jaws[2].

Cutting forces imposed by machining operations vary in magnitude and direction. A drilling operation induces *torque*, while a shaping operation causes straight-line *thrust*. The workholder must support the workpiece in opposition to the cutting forces and will generally be designed for a specific machining operation.

Many workholders are used in industry that are not used on material removing operations. Workholders may be used for the inspection of workpieces, assembly, welding, and so on. There may be very little difference in their basic design and their appearance. Quite often a standard commercial design may be used in one application for a turning operation and for the same or another workpiece in an inspection operation.

Questions

1. What is the purpose and function of the workholder?
2. Describe the principles of workholders.
3. What are the influence factors in design or selection of a workholder?

New Words and Expressions

1. workholder/ˈwəːkhəuldə/n. 工件夹紧装置
2. grip/grip/n. 紧握，啮合，柄；vt. & vi. 紧握，控制
3. chuck/tʃʌk/n. 夹头，夹盘，卡盘；vt. 夹紧，卡紧
4. pneumatic/njuːˈmætik/a. 空气的，气动的；n. 气胎
5. pliers/ˈplaiəz/n. 钳，手钳，台虎钳
6. tong/tɔŋ/n. （常用 tongs）钳，夹子；v. 用钳夹住
7. serration/seˈreiʃən/n. 锯齿状，成锯齿形
8. slippage/ˈslipidʒ/n. 滑动，打滑，空转，下降
9. vise/vais/n. 台虎钳；vt. 钳住，夹紧（= vice）
10. jaw/dʒɔː/n. 台虎钳牙，夹片，口部
11. attachment/əˈtætʃmənt/n. 连接物，附件，附加装置
12. depict/diˈpikt/vt. 描写，描述，描绘
13. wrench/rentʃ/n. 拧，扳钳，扳手；vt. & vi. 拧，扳紧
14. unison/ˈjuːnizn/n. 一致，统一
15. firmness/ˈfəːmnis/n. 坚固，坚定，稳固
16. abrupt/əˈbrʌpt/a. 突然的，陡峭的
17. torque/tɔːk/n. 转矩，扭矩，扭转
18. thrust/θrʌst/v. 猛推，冲，延伸；n. 拉力，推力，牵引力
19. except for 除了，只有
20. in contact with 和……接触着
21. tool-holder 刀夹，刀杆，刀柄
22. tool-holding 刀具夹紧
23. attachment screw 止动联接螺钉

Notes

[1] The vise remains locked by the self-locking characteristic of the screw, provides means of attachment to a machine, and permits precise placement of the work.

台虎钳是由螺旋的自锁特性来保持锁紧的，它提供了使其他部件附着到机床上的手段，从而确保加工时的精确定位。

句中 remains, provides, permits 为并列谓语，进一步说明台虎钳的作用。

[2] The workholder material must be carefully selected with the workpiece in mind so that neither will be damaged by abrupt contact, e. g., damage to a soft copper workpiece by hard steel jaws.

在考虑工件材料的前提下，必须细心选择夹具的材料，只有这样才不会引起接触性破坏。例如：若选用硬的钢质台虎钳口，就会使比较软的铜质工件材料受到破坏。

句中 so that neither will be damaged by abrupt contact 为目的状语从句。

Glossary of Terms

1. workpiece   工件，冲压件
2. holder of punch   凸模夹持器
3. locating device   定位装置
4. locating face   定位面
5. locating pin   定位销（挡料销）
6. locating plate   定位板
7. locating ring   定位圈
8. locating rule   定位尺
9. locating element   定位零件（定位要素）
10. workholding   工件夹紧
11. work hardening   加工硬化
12. internal cylindrical grinding machine   内圆磨床
13. internal cylindrical turning   内圆车削
14. internal force   内力
15. internal cylindrical grinding machine with vertical spindle   立式内圆磨床
16. hole scraping（turning, milling, lapping）刮孔（车孔，铣孔，研孔）
17. hole grinding（slotting, honing, flanging）磨孔（插孔，珩孔，翻孔）
18. versatile grinding machine   多用磨床
19. versatile lathe   多用车床
20. vertical multi-tool lathe   立式多刀车床
21. grip device（clamping device）夹紧装置
22. grip holder   夹头
23. precision milling machine   精密铣床

Reading Materials

Locating Principles

To insure successful operation of a workholding device, the workpiece must be accurately located to establish a definite relationship between the cutting tool and some points or surfaces of the workpiece. This relationship is established by locators in the workholding device which position and restrict the workpiece to prevent its moving from its predetermined location. The workholding device will then present the workpiece to the cutting tool in the required relationship. The locating device should be so designed that each successive workpiece, when loaded and clamped, will occupy the same position in the workholding device. Various methods have been devised to effectively restrict the movement of workpieces. The locating design selected for a given workholding device will depend on the nature of the workpiece, the requirements of the metal-removing operation to be performed, and other restrictions on the workholding device.

Types of Location

Basic workpiece location can be divided into three fundamental categories: plane, concentric, and radial. In many cases, more than one category of location may be used to locate a particular workpiece. However, for the purpose of identification and explanation, each will be discussed individually.

Plane location is normally considered the act, or process, of locating a flat surface. But many times irregular surfaces may also be located in this manner. Plane location is simply locating a workpiece with reference to a particular surface or plane (Figure 6-6).

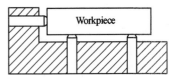

Figure 6-6   Plane location

Concentric location is the process of locating a workpiece from an internal or external diameter (Figure 6-7).

Radial location is normally a supplement to concentric location. With radial location (Figure 6-8), the workpiece is first located concentrically and then a specific point on the workpiece is located to provide a specific fixed relationship to the concentric locator.

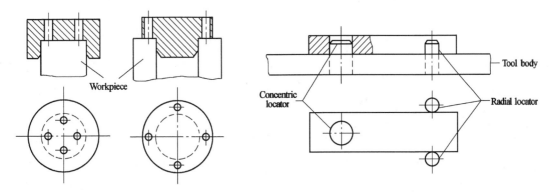

Figure 6-7   Concentric location   Figure 6-8   Radial location

Most workholders use a combination of locational methods to completely locate a workpiece. The part shown in Figure 6-9 is an example of all three basic types of location being used to reference a workpiece.

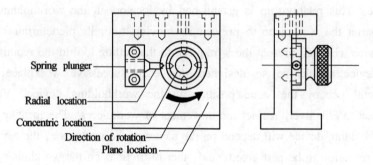

Spring plunger
Radial location
Concentric location
Direction of rotation
Plane location

Figure 6-9　Plane, concentric, and radial location

Lesson 7　Jig and Fixture Design

Text

Jigs are workholders which are designed to *hold*, *locate*, and support a workpiece while guiding the cutting tool throughout its cutting cycle. Jigs can be divided into two general classifications: drill jigs and boring jigs. Of these, drill jigs are, by far, the most common. Drill jigs are generally used for drilling, *tapping*, and *reaming*, but may also be used for *countersinking*, *counterboring*, *chamfering*, and *spotfacing*. Boring jigs, on the other hand, are normally used exclusively for boring holes to a *precise*, *predetermined* size. The basic design of both classes of jigs is essentially the same. The only major difference is that boring jigs are normally fitted with a *pilot bushing* or bearing to support the outer end of the boring bar during the machining operation.

In designing any jig, there are numerous considerations that must be addressed. Although several of these points, such as locating, supporting, and clamping, have already been covered, they are included in this section because they apply to jig design. Since all jigs have a similar construction, the points covered for one type of jig normally apply to the other types as well. Jig design and selection begins with an analysis of the workpiece and the manufacturing operation to be performed.

One of the first considerations in the design of any workholder is the relative balance between the cost of the tool and the expected *benefits* of using the tool for production[1]. All workholders should save more in production costs than the tool costs to design and construct. In many instances, tool designers may have to complete detailed estimates to justify the cost of special tooling. This involves a close look at the part drawing, process specifications and other related documents.

Typically, the complexity of the part, location and number of holes, required accuracy, and the number of parts to be made are all points which must be considered to determine if the cost of a particular jig is warranted. Once the tool designer is satisfied that the cost of special tooling is justified, the remaining data required to produce a suitable workholder is *compiled* and analyzed.

Fixtures are workholders which are designed to hold, locate, and support the workpiece during the machining cycle. Unlike jigs, fixtures do not guide the cutting tool, but rather provide a means to reference and *align* the cutting tool to the workpiece. Fixtures are normally classified by the machine with which they are designed to be used. A sub-classification is sometimes added to further specify the fixture classification. This sub-classification identifies the specific type of machining operation the fixture is intended to perform. For example: a fixture used with a milling machine is called a milling fixture, however, if the operation it is to perform is *gang* milling, it may also be called a gang-milling fixture. Likewise, a *bandsawing* fixture designed for *slotting* operations may also be referred to as a bandsaw-slotting fixture.

The similarity between jigs and fixtures normally ends with the design of the tool body. For the most part, fixtures are designed to withstand much greater stresses and tool forces than jigs, and are always securely clamped to the machine[2]. For these reasons, the designer must always be aware of proper locating, supporting, and clamping methods when fixturing any part.

In designing any fixture, there are several considerations in addition to the part which must be addressed to complete a successful design. Cost, production capabilities, production processing, and tool longevity are some of the points which must *share* attention with the workpiece when a fixture is designed.

As with all tooling, the first consideration in fixture design is the cost *versus* the benefit. The production quantity, rate, or accuracy must *warrant* the added expense of special tooling. In addition, the fixture must pay for itself with savings derived from its use in as short a time as possible.

Questions

1. What is a jig?
2. What are the functions of a fixture?
3. What is the difference between jig and fixture?

New Words and Expressions

1. jig/dʒig/ *n.* 夹具，模具，规尺，机架
2. hold/həuld/ *vt. & n.* 拿，握，支持，夹住
3. locate/ləu'keit/ *vt.* 设置，安排，定位，确定（零件等）位置
4. tapping/'tæpiŋ/ *n.* 螺纹加工，攻螺纹
5. reaming/'riːmiŋ/ *n.* 铰孔加工，扩孔
6. countersink/'kauntəsiŋk/ *n.* 埋头孔，锥口孔；*vt.* 钻（孔），打埋头孔于
7. counterbore/'kauntəbɔː/ *vt.* 扩孔，锪孔
8. chamfer/'tʃæmfə/ *n.* 槽，斜面，圆角，倒角；*vt.* 去角，斜切
9. spotfacing/'spɔt'feisiŋ/ *n.* 锪孔
10. precise/pri'sais/ *a.* 正确的，精确的，精密的
11. predetermine/'priːdi'dəːmin/ *vt.* 预定
12. pilot/'pailət/ *n.* 领航员，定料销，导向器
13. bush/buʃ/ *n.* 衬套，轴瓦；*vt.* 加衬套于……
14. benefit/'benifit/ *n.* 利益，好处；*vt.* 有益于……；*vi.* 受益
15. compile/kəm'pail/ *vt.* 编纂，编辑，编译程序
16. fixture/'fikstʃə/ *n.* 夹具，夹紧装置，型架，固定
17. align/ə'lain/ *vt.* 匹配，排列成一行，定位，校直
18. gang/gæŋ/ *n.* 一组，一队，帮，伙；*v.* 连接
19. bandsaw/'bændsɔː/ *n.* 带锯，带锯机
20. slot/slɔt/ *n.* 切口，裂口，槽沟；*vt.* 开槽于
21. share/ʃɛə/ *n.* 一份，股份；*vt.* 均分，分配
22. versus/'vəːsəs/ *prep.* ……对……，与……比较，依……为转移
23. warrant/'wɔrənt/ *n.* 证明，理由，根据；*vt.* 保证，批准
24. slotter/'slɔtə/ *n.* 插床，侧床，立刨床
25. spindle/'spindl/ *n.* 轴，主轴，杆，蜗杆
26. gang-milling 多刀铣削，排铣
27. apply to 致力于，适用于
28. (be) aware of 知道，意识到，认识

Notes

［1］One of the first considerations in the design of any workholder is the relative balance between the cost of the tool and the expected benefits of using the tool for production.

在设计任何夹具时，首先要考虑的问题是制造该工具的成本和使用该工具进行生产所希望产生的效益两者之间应保持相对平衡。

单个分词 expected 作名词 benefits 的前置定语；动名词短语 using the tool for production作介词 of 的宾语。

［2］For the most part, fixtures are designed to withstand much greater stresses and tool forces than jigs, and are always securely clamped to the machine.

就大多数情况而言，fixtures（夹具）在设计上比 jigs（钻模）能够承受更大的应力，并且总能可靠地将工件夹紧到机床上。

介词短语 for the most part 在这里相当于 in most cases，可译为"在大多数情况下"。

Glossary of Terms

1. spot face　孔口平面
2. drill and countersink　定心钻，中心钻
3. counterbore cutter head　扩孔钻头
4. jig boring machine　坐标镗床
5. jig grinding machine　坐标磨床
6. jig and fixture　夹具
7. fixture of gear cutting machine　齿轮加工机床夹具
8. fixture of grinding machine　磨床夹具
9. fixture of milling machine　铣床夹具
10. fixture of planing machine　刨床夹具
11. fixture of slotting machine　插床夹具
12. vacuum fixture　真空夹具
13. universal fixture（jig）　通用夹具
14. stationary fixture　固定夹具
15. standard fixture（jig）　标准夹具
16. pneumatic fixture（jig）　气动夹具
17. open-side boring and milling machine　悬臂镗铣床
18. magnetic fixture（jig）　磁力夹具
19. locating device（face, element）　定位装置（面，元素）
20. hydraulic fixture（jig）　液压夹具

Reading Materials

Types of Jigs

Jigs are made in a wide variety of different styles. The specific style or type of jig which should be used for a particular application is generally determined by the workpiece itself. The two general categories of jigs are open and closed. Open jigs are normally used for parts which only require machining on a single surface, or side. Closed jigs are used to machine parts which require operations on more than one side or surface. The terms commonly used to describe specific types of jigs within each category are generally associated with the basic appearance or construction of the jig.

Template Jigs. Template jigs are generally flat, open jigs which are used to locate the general position of drilled holes. This type of jig does not normally contain any device for clamping or securing the jig to the workpiece, but relies on auxiliary clamps to hold the jig when necessary. Template jigs are the least expensive type of jig, and are frequently used where extreme accuracy is not required. When used for limited production, template jigs do not normally contain drill bushings. Instead, the entire jig plate is hardened. Figure 7-1 shows several examples of applications for which template jigs are well suited.

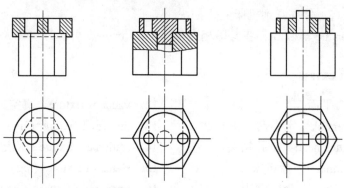

Figure 7-1　Applications for template jigs

Plate Jigs. Plate jigs are a somewhat more sophisticated variation of the basic template jig. These jigs are generally more accurate and durable than template jigs and also include a means to securely clamp the workpiece. The principal variations of the plate jig commonly used for stationary workholders are the plain plate jig, table jig, and the sandwich jig.

Plain Plate Jigs. Plain plate jigs (Figure 7-2) consist of a jig plate which forms the main body of the jig and contains the drill bushings (when used), locators, and clamping device. Depending on the complexity of the workpiece, it is often less expensive to make several plate jigs than a single box, or closed jig.

Table Jigs. Table jigs (Figure 7-3) are another variation of the plate jig which uses the jig

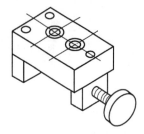

Figure 7-2　Plain plate jig

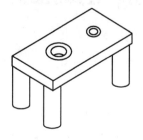

Figure 7-3　Table jig

plate as a main member with the other components attached. The principal difference is the addition of the legs which raise the jig off the machine table.

Sandwich Jigs. Sandwich jigs (Figure 7-4) consist of two plates which are used to sandwich the workpiece. These jigs use a jig plate to establish the location of the drill bushing and the position of the part to be drilled.

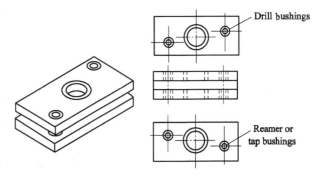

Figure 7-4 Sandwich jig

Types of Fixtures

Fixtures are classified either by the machine they are used on, or by the process they perform on a particular machine tool. However, fixtures also may be identified by their basic construction features. For example, a lathe fixture made to turn radii is classified as a lathe radius turning fixture. But if this same fixture were a simple plate with a variety of locators and clamps mounted on a faceplate, it is also a plate fixture. Like jigs, fixtures are made in a variety of different forms. While many fixtures use a combination of different features, almost all can be divided into five distinct groups. These include plate fixtures, angle plate fixtures, vise jaw fixtures, indexing fixtures, and multi-part, or multi-station fixtures.

Plate fixtures, as their name implies, are constructed from a plate with a variety of locators, supports, and clamps (Figure 7-5). Plate fixtures are the most common type of fixture. Their versatility makes them adaptable for a wide range of different machine tools. Plate fixtures may be made from any number of different materials, depending on the application of the fixture.

The angle plate fixture (Figure 7-6) is a modified form of plate fixture. Here, rather than having a reference surface parallel to the mounting surface, the angle plate fixture has a reference surface perpendicular to its mounting surface. This construction is very useful for those machining operations which are performed perpendicular to the primary reference surface of the fixture.

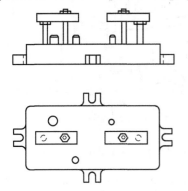

Figure 7-5 Plate fixture

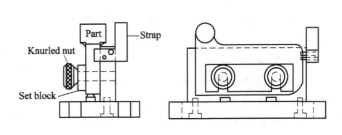

Figure 7-6 Angle plate fixture

Unit Three

Lesson 8   Press Types

Text

Characteristic of the press working process is the application of large forces by press tools for a short time *interval*, which results in the cutting (shearing) or *deformation* of the work material.

A pressworking operation, generally completed by a single application of pressure, often results in the production of a finished part in less than one second.

Pressworking forces are set up, guided, and controlled in a machine referred to as a press.

Power Presses. Essentially, a press is comprised of a frame, a bed or *bolster* plate, and a *reciprocating* member called a *ram* or *slide* which exerts force upon work material through special tools mounted on the ram and bed.

Energy stored in the rotating *flywheel* of a mechanical press (or supplied by a *hydraulic* system in a hydraulic press) is transferred to the ram for its linear movements.

Press Types. An open-back inclinable (OBI) press (Figure 8-1), also called a gap-frame press, has a C-shaped frame which allows access to its working space (between the bed and the ram). The frame can be *inclined* at an angle to the base, allowing for the *disposal* of finished parts by gravity. The open back allows the feeding and unloading of *stock*, workpieces, and finished parts through it from front to back.

Major components of a press are as follows:

1. **Press Bed.** A *rectangular* part of the frame, generally open in its center, which supports a bolster plate.

2. **Bolster Plate.** A flat steel plate,

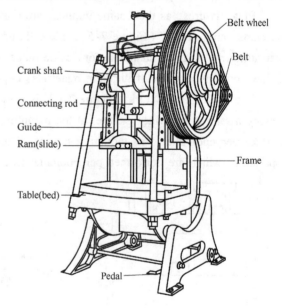

Figure 8-1   Open-back inclinable press

from 2 to 5in (51 ~ 127mm) thick, upon which press tools and accessories are mounted. Bolsters having standard dimensions and openings are available from press manufacturers.

3. **Ram or Slide.** The upper press member that moves through a *stroke* a distance depending upon the size and design of the press. The position of the ram, but not its stroke, can be adjusted.

The distance from the top of the bed (or bolster) to the bottom of the slide, with its stroke down and adjustment up, is the *shut* height of a press[1].

4. *Knockout.* A mechanism operating on the upstroke of a press, which *ejects* workpieces or blanks from a press tool.

5. *Cushion.* A press *accessory* located beneath or within a bolster for producing an upward motion and force; it is actuated by air, oil, rubber, springs or a combination of mechanisms.

A straight side press of conventional design has columns (uprights) at the ends of the bed, usually with windows (square or rectangular openings) to allow the feeding and unloading of stock, workpieces, and finished parts.

With special applications, this type of press also can be used for feeding from front to back (Figure 8-2).

A press brake is essentially the same as a *gap*-frame press except for its long bed from 6 to 20 feet (1.8~6m) or more. It is used basically for various bending operations on large sheet metal parts. It can also be used with a series of separate sets of press tools to do light piercing, *notching*, and forming. This allows parts of a complex design to be accurately made without a high-cost press tool by simply breaking the complex part down into several simple operations. This type of operation is used on low-run or *prototype* parts. The tooling cost is usually very low, but the labor cost is high as the parts are manually transferred and located in each station. The operator must follow good safety practices at all times to avoid *injury*.

A hydraulic press is used basically for forming operations and a slower operating cycle time than most mechanical presses. The advantages of hydraulic presses are that the working pressure, stroke, and speed of the ram are adjustable (Figure 8-3).

A double-action press is used for large, or deep drawing operations on sheet metal parts. This type of press has an outer ram (blank holder) and a second inner ram (punch holder). During the operating cycle, the blank holder contacts the material first and applies pressure to allow the punch holder to properly draw the part.

A *triple*-action press has the same inner and outer ram as the double-action press, but a third ram in the press bed moves up allowing a *reverse* draw to be made in one press cycle[2]. The triple-action press is not widely used.

A *knuckle* press is used for *coining* operations. The design of the drive allows for very high pressures at the bottom of the ram stroke. This type uses a *crank*, which moves a joint consisting of two levers that *oscillate* to and from dead center and results in a short, powerful movement of the slide with slow travel near the bottom of the stroke.

These are the basic press types used in industry, although there are many more types with special applications.

Press operator safety must be a primary concern for everyone in the press area. While working under the ram the press control must be locked in the off position and safety blocks placed under the ram to prevent it from coasting down. While the press is running, the use of proper guards and safety procedures must always be followed.

Press safety under OHSA (Occupational Health and Safety Act) is law. ***Strict compliance*** to the regulations are required.

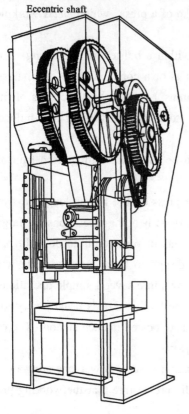

Figure 8-2　Single-action, straight-side,
eccentric-shaft mechanical press

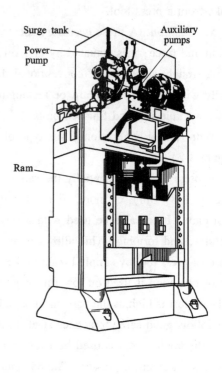

Figure 8-3　Typical hydraulic press

Questions

1. List the six types of presses.
2. What are major components of a press?
3. Describe an open-back inclinable press.
4. Press operator safety must be a primary concern for everyone in the press area, why?

New Words and Expressions

1. interval/ˈintəvəl/ *n.* 间隔，空间，周期
2. deformation/difɔːˈmeiʃən/ *n.* 形变，畸变，失真
3. bolster/ˈbəulstə/ *n.* 垫枕，垫木，垫枕状的支撑物；*vt.* 支撑，垫，支持
4. reciprocate/riˈsiprəkeit/ *vt.* 使往复运动，互换，互给
5. ram/ræm/ *n.* 压头，柱塞，滑枕，撞

杆；*vt.* 撞击，冲压

6. slide/slaid/*vi.* 滑，溜；*vt.* 使滑入；*n.* 滑动，滑块，滑座，冲头

7. flywheel/ˈflaihwiːl/*n.* 飞轮，惯性轮，整速轮

8. hydraulic/haiˈdrɔːlik/*a.* 水力（学）的，液压的，液力的

9. incline/inˈklain/*vt.* 倾斜，偏向；*n.* 斜坡，斜面

10. disposal/disˈpəuzəl/*n.* 安排，排列，处理，配置

11. stock/stɔk/*n.* 杠杆，钻柄，原料，材料，台，座；*vt.* 给……装上把手

12. rectangular/rekˈtæŋgjulə/*a.* 长方形的，矩形的，成直角的

13. stroke/strəuk/*n.* 打，击，冲程，行程

14. shut/ʃʌt/*vt.* 关闭，封闭，关上，合拢；*n. vi.* 关闭

15. knock/nɔk/*vt.* 打击，碰撞；*n.* 敲打

16. eject/iˈdʒekt/*vt.* 顶出，抛出，排斥

17. cushion/ˈkuʃən/*n.* 垫子，气垫，缓冲器；*vt.* 使减少震动

18. accessory/ækˈsesəri/*n.* 附件，附属设备；*a.* 附加的，次要的

19. gap/gæp/*n.* 裂口，缺口，开口，间隙；*vt.* 使成缺口

20. notch/ˈnɔtʃ/*n.* 切口，切痕；*v.* 开缺口，冲缺口

21. prototype/ˈprəutətaip/*n.* 原型，模型，样机，样品

22. injury/ˈindʒəri/*n.* 损害，伤害

23. triple/ˈtripl/*a.* 三倍的，三重的，三联的；*v.* 使增加三倍

24. reverse/riˈvəːs/*vt.* 颠倒，反转；*vi.* 倒退，反向；*n.* 反向，倒退；*a.* 反面的

25. knuckle/ˈnʌkl/*n.* 万向接头，指节，肘节

26. coining/ˈkɔiniŋ/*n.* 压印加工，压花，立体挤压

27. crank/kræŋk/*n.* 曲柄，曲轴，手柄，弯曲

28. oscillate/ˈɔsileit/*vi.* 波动，振动；*vt.* 使摆动

29. strict/strikt/*a.* 严格的，精确的，严密的

30. compliance/kəmˈplaiəns/*n.* 答应，服从，可塑性，配合性，柔量

31. result in 结果形成，导致

32. a series of 一系列，许多

33. (be) comprised of 包括在……内

34. triple screw 三线螺旋，三纹螺旋

35. prototype workpiece 样件

Notes

[1] The distance from the top of the bed (or bolster) to the bottom of the slide, with its stroke down and adjustment up, is the shut height of a press.

从工作台（或垫板）的上表面到滑块的下表面之间的距离（可以上、下调节），称为压力机的闭合高度。

句中介词短语 from the top of the bed (or bolster) to the bottom of the slide 属于 "from. . . to. . ."，可译为"从……到……"，这个介词短语作后置定语，修饰 the distance。

[2] A triple-action press has the same inner and outer ram as the double-action press, but a third ram in the press bed moves up allowing a reverse draw to be made in one press cycle.

三动压力机具有和双动压力机相同的内、外滑块。此外，三动压力机工作缸还有另一个滑块，它可向上运动，从而在一个冲压循环中允许反向拉深。

句中短语 the same inner and outer ram as the double-action press 属于 the same... as... 这种结构；inner and outer ram 译为 "内、外滑块"。原文中的 "a third" 主要侧重于 "another（另一个）" 的意思，而不强调次序。

Glossary of Terms

1. slide gauge 游标卡尺
2. slide rule 计算尺
3. die slide 下模滑动装置
4. triple action press 三动压力机
5. turret press 冲模回转压力机
6. two point press 双点压力机
7. twin-drive press 双边齿轮驱动压力机
8. two point single action press 双点单动压力机
9. bench press 台式压力机
10. trimming press 切边压力机
11. closed type single action crank press 闭式单动（曲柄）压力机
12. knuckle joint press 肘杆压力机
13. one point single action press 单点单动压力机
14. open-back inclinable press 开式双柱可倾压力机
15. open side press 开式压力机，单柱压力机
16. hydroformer press 液压成形压力机
17. four point press 四点压力机
18. four crank press 四曲柄压力机
19. flywheel-type screw press 飞轮式螺旋压力机
20. friction screw press 摩擦（传动）螺旋压力机
21. single action double crank press 单动双曲柄压力机
22. single piece frame press 整体框架式压力机
23. rocker arm type press 摇臂式压力机
24. punch press 冲床，（冲裁）压力机
25. top drive sheet metal stamping automatic press 上传动板料冲压自动压力机
26. hydraulic press 液压机
27. mechanical（power）press 机械压力机
28. screw press 螺旋压力机
29. drive screw 传动螺杆

Reading Materials

Screw Presses

Screw presses are energy-constrained forming machines which obtain their work capacity from a rotating mass (flywheel). Figure 8-4 shows a variety of flywheel drives. The rotary motion of the flywheel is converted into the linear motion of the ram in the machine frame by the use of a lead screw and nut drive. The drive for the flywheel is normally arranged with its electric motor (Figure 8-4a). The flywheel has a fixed connection with the lead screw, which usually has a three-start thread. The nut is fixed at the head of the machine frame and the lead screw, together with its flywheel, travels in the nut in a vertical direction. During the downward motion, the radius of contact of the driving disc increases. A desirable increase in flywheel speed results, although the driving disc rotates at constant speed. On the return movement of the ram, the opposite driving disc is in contact with the flywheel, but the transmission relationships are disadvantageous. While the contact is at the maximum diameter of the driving disc, the flywheel must be accelerated from a near-stationary state. Consequently, there is a great deal of slipping and consequent wear of the friction surfaces. With more-complicated drives, e. g. those shown in Figure 8-4b and Figure 8-4c, this disadvantage of the conventional triple-disc drive is avoided.

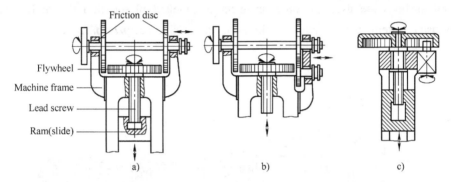

Figure 8-4　Flywheel drives for screw presses
a) Triple-disc drive　b) Four-disc drive　c) Single-disc drive

Hydraulic Presses

Hydraulic presses belong to the force-constrained type of forming machines. Their main use is found in those areas of forming technology where the force along the path of the ram must remain constant or under accurate control. The drive mechanism of piston and cylinder acts in a linear manner and is directly connected to the ram. The form of frame construction of hydraulic presses is largely similar to that of mechanical presses. The hydraulic drive units are easily accommodated in the machine frame. Consequently, several hydraulic drives can readily be built into a single machine for complicated forming and cutting operations (drawing, extruding, cutting, swaging, etc.),and the required motions may be easily coordinated.

Lesson 9　The Injection-Molding Machines

Text

The greatest quantities of plastic parts are made by injection molding. The process consists of feeding a plastic **compound** in **powdered** or **granular** form from a **hopper** through **metering** and melting stages and then injecting it into a mold[1]. After a brief cooling period, the mold is opened and the solidified part ejected. In most cases, it is ready for immediate use.

Several methods are used to force or inject the melted plastic into the mold. The most commonly used system in the larger machines is the in-line **reciprocating** screw, as shown in Figure 9-1.

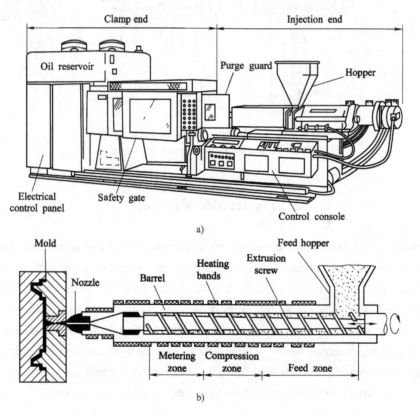

Figure 9-1　The injection-molding machine and injection system

a) The injection-molding machine　b) The reciprocating-screw injection system

The screw acts as a combination injection and plasticizing unit. As the plastic is fed to the rotating screw, it passes through three zones as shown: feed, compression, and metering. After the feed zone, the screw-flight depth is gradually reduced, forcing the plastic to compress. The work is converted to heat by shearing the plastic, making it a semifluid mass. In the metering zone, additional heat is applied by conduction from the **barrel** surface. As the chamber in front of

44

the screw becomes filled, it forces the screw back, ***tripping*** a limit switch that ***activates*** a hydraulic cylinder that forces the screw forward and injects the fluid plastic into the closed mold[2]. An antiflowback valve prevents plastic under pressure from escaping back into the screw flights.

The clamping force that a machine is capable of exerting is part of the size designation and is measured in tons. A rule-of-thumb can be used to determine the tonnage required for a particular job. It is based on two tons of clamp force per square inch of projected area. If the flow pattern is difficult and the parts are thin, this may have to go to three or four tons.

Many reciprocating-screw machines are capable of handling thermosetting plastic materials. Previously these materials were handled by compression or transfer molding. Thermosetting materials cure or ***polymerize*** in the mold and are ejected hot in the range of 375 ~ 410°F (190 ~ 210°C). Thermoplastic parts must be allowed to cool in the mold in order to remove them without distortion. Thus thermosetting cycles can be faster. Of course the mold must be heated rather than chilled, as with thermoplastics.

Ways of injection molding plastic material are ***sketched*** in Figure 9-2. The oldest is the single-stage plunger method. When the plunger is drawn back, raw material falls from the hopper into the chamber. The plunger is driven forward to force the material through the heating cylinder where it is softened and ***squirted*** under pressure into the mold. The single-stage reciprocating screw system has become more popular because it prepares the material more thoroughly for the

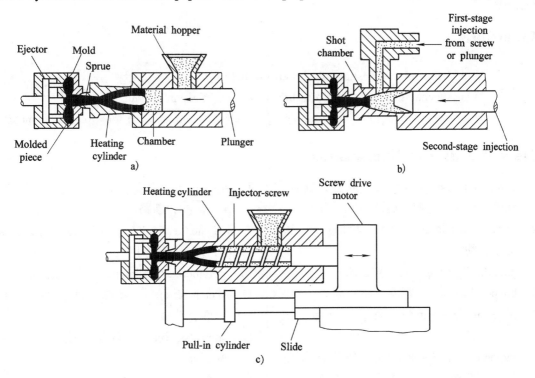

Figure 9-2　Injection molding systems

a) Conventional single-stage plunger type　b) Two-stage plunger or screw-plasticizer types

c) Single-stage reciprocating screw type

mold and is generally faster. As the screw turns, it is pushed backward and *crams* the charge from the hopper into the heating cylinder. When enough material has been prepared, the screw stops turning and is driven forward as a plunger to ram the charge into the die. In a two-stage system, the material is plasticized in one cylinder, and a definite amount transferred by a plunger or screw into a shot chamber from which a plunger injects it into the mold.

An injection molding machine heats to soften, molds, and cools to harden a thermoplastic material. Operating-temperature is generally between 150℃ and 380℃ (300℉ and 700℉) with full pressure usually over 35 and up to 350 MPa (5000 to 50000 psi). The mold is water cooled. The molded piece and sprue are *withdrawn* from the injection side and ejected from the other side when the mold is opened. The mold is then closed and clamped to start another cycle. Thermosetting plastics can be injection molded but have to be polymerized and molded before they set in the machine. This may be done in a reciprocating screw machine where one charge at a time is brought to curing temperature. By another method, sometimes called jet molding, preforms are charged one at a time into a single-stage plunger machine.

Machines are available for molding sandwich parts. One cylinder and plunger injects a measured amount of skin material into the die, and then a second cylinder squirts the filler inside the mass. Finally, a final *spurt* from the first cylinder clears the core material from the sprue. The aim is to produce composites with *optimum* properties. Either case or core may be *foamed*.

Questions

1. What happens if a plastic mold is run too hot?
2. What mechanism creates the pressure for plastic injection molding?
3. How does one go about determining the required clamping force for a particular mold?

New Words and Expressions

1. compound/kəm'paund/*vt.* 混合，复合，组合；*n.* 化合物，混合物
2. powder/'paudə/*n.* 粉末，粉剂；*v.* 磨成粉，粉化
3. granular/'grænjulə/*a.* 粒状的
4. hopper/'hɔpə/*n.* 漏斗，给料斗
5. metering/'miːtəriŋ/*n.* 测量，计量，记录，统计
6. reciprocate/ri'siprəkeit/*vt.* 使机件往复运动，互换；*vi.* 往复移动，互换
7. barrel/'bærəl/*n.* 圆筒，桶，圆柱体，燃烧室；*vt.* 把……装桶
8. trip/trip/*n.* 往返，行程，脱开，切断；*vt.* 解扣，断路
9. activate/'æktiveit/*vt.* 使活动，开动，起动，对……起作用
10. polymerize/'pɔliməraiz/*v.* 使聚合
11. sketch/sketʃ/*n.* 草图，设计图；*v.* 画草图，草拟
12. squirt/skwəːt/*v.* 喷湿，喷（出）；*n.* 喷射器
13. cram/kræm/*n.* 填塞，压碎；*vt.* 塞入，塞满
14. withdraw/wið'drɔː/*v.* 撤回，缩回，取

消，拉开，移开

15. spurt/spə:t/v. & n. 喷出，进出，溅散

16. optimum/ˈɔptiməm/n. 最佳条件，最佳状况，最佳值；a. 最合适的

17. foam/fəum/n. 泡沫，泡沫材料，泡沫塑料；v. （使）起泡沫

18. （be）based on 以……作为……的根据

19. （be）capable of 能……的，易……的

Notes

[1] The process consists of feeding a plastic compound in powdered or granular form from a hopper through metering and melting stages and then injecting it into a mold.

注射过程（工艺）包括两个阶段，一是经料斗送入的粉状或粒状形式的塑料混合物通过熔融和定量区，二是将熔融塑料注射到型腔中。

句中 consist of 可译为："由……组成，包括"，from... into... 可译为："从……到……"。

[2] As the chamber in front of the screw becomes filled, it forces the screw back, tripping a limit switch that activates a hydraulic cylinder that forces the screw forward and injects the fluid plastic into the closed mold.

当螺杆前部的加料室被填满塑料时，就会迫使螺杆后退，进而断开限位开关而开动液压缸，从而使螺杆向前运动，将熔融塑料注射到闭合的模具型腔中。

句中 that activates a hydraulic cylinder that forces the screw forward and injects the fluid plastic into the closed mold 为限定性定语从句，修饰 limit switch；而 that forces the screw forward and injects the fluid plastic into the closed mold 又修饰 hydraulic cylinder。

Glossary of Terms

1. barrel surface 圆柱形表面
2. antiflowback valve 止逆阀
3. reciprocating-screw machine 往复螺杆式注射机
4. single-stage plunger 单级柱塞
5. shot chamber 注射室
6. curing temperature 固化温度
7. metering zone 定量段
8. metering jet 量嘴，量射口
9. shot volume 注射量，压注量
10. shot cylinder 压射缸
11. shot capacity 注射能力
12. injection piston 压射冲头
13. injection pressure 压射比压，注射压力
14. injection speed 注射速度
15. injection forming 注射成型
16. screw feeder 螺旋给料机
17. plunger diameter 柱塞直径
18. velocity of plunger 柱塞速度
19. sectional area of plunger 柱塞面积
20. hydraulic cylinder 液压缸

Reading Materials

2-Platen Injection Molding Machines

Design Advantages

1. Advanced, reliable, rugged direct acting clamp
2. Low profile machine design
3. Energy efficient, variable volume pump with proportional hydraulics
4. Controlled-stress tie rods
5. Clamp traversing cylinders
6. Generous mold space for production versatility
7. Full time oil filtration with indicator
8. Powered operator gate (1500 tons and larger)
9. Efficient, precise twin cylinder injection units
10. Interchangeable screw/barrel combinations for full process versatility
11. Twin pull-in cylinders for injection unit

Injection Features

1. 5-stage injection pressure: (1) injection high, (2) pack, (3) hold
2. 5-stage profiled injection velocities
3. 5-stage back pressure control
4. 2-stage screw rpm
5. PID temperature control of nozzle and barrel
6. Slide shutoff on hopper
7. Cold screw start protection
8. Injection transfer on position, hydraulic pressure, or time
9. Multiple screw and barrel combination
10. 27500 psi injection pressure on "A" barrels
11. Hopper discharge chute
12. Ball check or slider screw tip
13. Bimetallic barrel (110 mm injection frame and larger)
14. Power injection unit swivel (110mm injection frame and larger)

Maxim1500 Specifications

Clamp Unit Specifications

Clamp Force /kN	13344	Maximum Mold Wt /kg	30620
Clamp Opening Force(@ 6%) /kN	800	Platen Size(H × V) /mm	2470 × 1960
Clamp Stroke /mm	2250	Distance Between Tie Rods(H × V) /mm	1840 × 1330
Clamp speed /mm · s^{-1}	760	Tie Rod Diameter /mm	250
Dry Cycle Time(typical) @ 50% stroke /s	7.0	Ejector Stroke(Max) /mm	350
Maximum Daylight /mm	2700	Ejector Force @ 1500 psi /kN	249
Minimum/Maximum Mold Thickness /mm	450/1500		

Machine Specifications (Dimension, Overall)

Length (with 100 mm lu) /mm	10760	Height /mm	2810
Width /mm	3845	Ship Wt /kg	69672

Water Requirements

Heat Exchanger @ 85 deg F /L · min^{-1}	170

Electric and Hydraulic

Machine Hyd. System Pressure, Max /bar[1]	190	Fixed Volume Capacity /L · min^{-1}	375
Hydraulic Pump Capacity @ 100 psi (total) / L · min^{-1}	750	Electric Motor / kW	112
Variable Volume Capacity /L · min^{-1}	261	Total Oil Reservoir Capacity /L	1893

[1] 1 bar = 10^5 Pa.

Injection Unit Specifications

	A	B	C
Injection Capacity, Max. G. P styrene /g	3288	3979	5138
Displacement /L	3.456	4.181	5.4
Injection Pressure Max. /bar	1897	1566	1214
Injection Rate /L · s^{-1}	1.016	1.245	1.606
Screw Stroke /mm	440	440	440
Screw Diameter /mm	100	110	125
Screw L/D ratio	22.0	20.0	17.6
Screw Performance			
Low Torque Screw Speed, Max /r · min^{-1}		165	
Low Torque at Screw /N · m		8265	
High Torque Screw Speed, Max /r · min^{-1}		106	
High Torque at Screw /N · m		12812	
Barrel Heat Control			
Number of Pyrometers (Barrel/Nozzle)		4/1	
Total Heat Capacity /kW		62	

Unit Four

Lesson 10　Blanking Technique

Text

Cutting (Shearing) Operations. In the following discussion, certain die *terminology* will be used *frequently*. Figure 10-1 presents the terms most commonly *encountered*.

Shear Action in Die Cutting Operations. The cutting of metal between die components is a shearing process in which the metal is *stressed* in shear between two cutting edges to the point of *fracture*, or *beyond* its ultimate strength.

The metal is subjected to both tensile and compressive stresses (Figure 10-2); *stretching* beyond the elastic limit occurs, then plastic deformation, reduction in area, and finally, fracturing starts through cleavage planes in the reduced area and becomes complete.

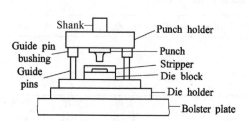

Figure 10-1　Common components of a simple die

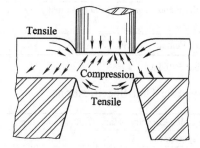

Figure 10-2　Stresses in die cutting

The fundamental steps in shearing or cutting are shown in Figure 10-3. The pressure applied by the punch on the metal tends to deform it into the die opening. When the elastic limit is exceeded by further loading, a portion of the metal will be forced into the die opening in the form of an *embossed* pad on the lower face of the material. A *corresponding* depression results on the upper face, as indicated at Figure 10-3a. As the load is further increased, the punch will penetrate the metal to a certain depth and force an equal portion of metal thickness into the die, as indicated at Figure 10-3b. This *penetration* occurs before fracturing starts and reduces the cross-sectional area of metal through which the cut is being made. Fracture will start in the reduced area at both upper and lower cutting *edges*, as indicated at Figure 10-3c. If the *clearance* is suitable for the material being cut, these fractures will spread toward each other and *eventually* meet, causing complete **separation**[1]. Further travel of the punch will carry the cut portion through the **stock** and into the die opening.

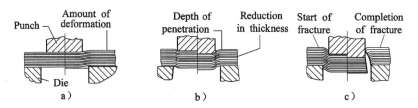

Figure 10-3   Steps in shearing metal
a) Plastic deformation   b) Reduction in thickness   c) Fracture

Center of Pressure. If the **contour** to be blanked is irregularly shaped, the **summation** of shearing forces on one side of the center of the ram may greatly exceed the forces on the other side. Such irregularity results in a bending moment in the press ram, and **undesirable deflections** and **misalignment**. It is therefore necessary to find a point about which the summation of shearing forces will be **symmetrical**. This point is called the center of pressure, and is the center of gravity of the line that is the **perimeter** of the blank. It is not the center of **gravity** of the area.

The press tool will be designed so that the center of pressure will be on the **axis** of the press ram when the tool is mounted in the press.

Clearances. Clearance is the space between the **mating** members of a die set. Proper clearances between cutting edges enable the fractures to meet. The fractured portion of the sheared edge will have a clean **appearance**. For **optimum** finish of a cut edge, proper clearance is necessary and is a function of the type, thickness, and temper of the work material. Clearance, penetration, and fracture are shown **schematically** in Figure 10-4. In Figure 10-5, characteristics of the cut edge on stock and blank, with normal clearance, are shown schematically. The upper corner of the cut edge of the stock (indicated by A′) and the lower corner of the blank (indicated by A′-1) will have a radius where the punch and die edges, **respectively**, make contact with the material. This radiusing is due to the plastic deformation taking place, and will be more **pronounced** when cutting soft metals[2]. Excessive clearance will cause a large radius at these corners, as well as a **burr** on opposite corners.

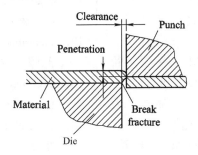

Figure 10-4   Punch-and-die clearance; punch penetration into and fracture of die-cut metal

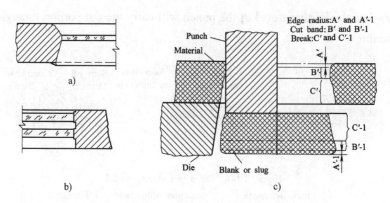

Figure 10-5　Cut-edge characteristics of die-cut metal；
effect of excessive and insufficient clearances

a）Excessive clearance　b）Insufficient clearance　c）Normal clearance

Questions

1.　What is the center of pressure?

2.　Describe shear action in die cutting operations.

3.　What is the fundamental steps in shearing or cutting?

4.　Why is it necessary for proper clearance in shearing or cutting?

5.　When the contour to be blanked is irregularly shaped, why must the center of pressure be calculated?

New Words and Expressions

1.　terminology /təːmiˈnɔlədʒi/ n.　术语，专门名词

2.　frequent /ˈfriːkwənt/ a.　频繁的，屡次的，常常的，习以为常的；vt.　常去，常到；-ly ad.　常常

3.　encounter /inˈkauntə/ vt.　遭遇，碰见，碰撞；vi.　偶遇；n.　遭遇

4.　stress /stres/ n.　压力，紧迫，应力；vt.　强调，使受应力

5.　fracture /ˈfræktʃə/ n.　断口，断面，断裂，破碎；vt.　断裂，折断，破碎

6.　beyond /biˈjɔnd/ prep.　在……那边，远于，迟于；超出；ad.　在远处

7.　stretch /stretʃ/ vt. vi. n.　伸展，展开，加宽

8.　emboss /imˈbɔs/ vt.　拷花，压纹，在……上浮雕图案

9.　correspond /kɔrisˈpɔnd/ vi.　相当，对应，符合

10.　penetration /peniˈtreiʃən/ n.　渗透，穿透，穿透能力

11.　edge /edʒ/ n.　刃，刀口，边缘，边线；vi.　使锐利，给……镶边

12.　clearance /ˈkliərəns/ n.　清除，消除，间隙，空隙

13.　eventual /iˈventjuəl/ a.　最后的，结局的，可能发生的，万一的

14.　separation /sepəˈreiʃən/ n.　分离，分类，间隔

15.　stock /stɔk/ n.　树干，托柄，原料，材

料，台，座；vt. 给……装上把手

16. contour/ˈkɔntuə/n. 轮廓，外形，等高线；vt. 描画轮廓线

17. summation/səˈmeiʃən/n. 总结，总和，总数，加法，求和

18. undesirable/ˌʌndiˈzaiərəbl/a. 不希望的，不合乎需要的

19. deflection/diˈflekʃən/n. 偏转，偏斜，偏移，偏差，挠曲，弯曲

20. misalignment/misəˈlainmənt/n. 未对准，不正，失调，不重合

21. symmetrical/siˈmetrikəl/a. 对称的，匀称的

22. perimeter/pəˈrimitə/n. 圆周，周长，周边，周界线

23. gravity/ˈgræviti/n. 重力，引力，重要性，严重性

24. axis/ˈæksis/n. (pl. axes) 轴，轴线，坐标轴，中心线

25. mate/meit/n. 同事，配对物，联接；v. 成配偶，紧密配合

26. appearance/əˈpiərəns/n. 出现，显露，外表，外貌

27. optimum/ˈɔptiməm/n. 最佳条件，最佳状况；a. 最适合的

28. schematic/skiˈmætik/a. 图解的，按照图示的，纲要的

29. respective/risˈpektiv/a. 各自的，各个的

30. pronounced/prəˈnaunst/a. 显著的，明显的，明确的

31. burr/bəː/n. 毛刺，毛边，轴环，套环，垫圈；v. 去毛刺

32. ultimate strength 极限强度

33. (be) subjected to 使受到，使遭遇

34. tensile and compressive stresses 拉和压应力

35. elastic limit 弹性极限

36. plastic deformation 塑性变形

37. cross-sectional area 横截面积

38. in many cases 在许多方面

39. spring-loaded stripper 弹性卸料板

Notes

[1] If the clearance is suitable for the material being cut, these fractures will spread toward each other and eventually meet, causing complete separation.

对于被剪切的材料，若间隙适当，则裂纹将相向扩展并最终相遇，从而使材料完全分离。

原文主句中有两个谓语动词，一个是 spread，另一个是 meet，它们共用一个主语，即 these fractures。分词短语 causing complete separation 用作状语，表示结果，汉译为"从而使材料完全分离"。注意：在科技文章中，分词短语作状语时，一般都用","把它与句子的其他部分分开。

[2] This radiusing is due to the plastic deformation taking place, and will be more pronounced when cutting soft metals.

这个圆角是由于材料发生塑性变形而引起的，并且在冲裁比较软的金属材料时圆角会更明显。

句中的 due to 意为"由于……原因"，分词短语 taking place 是 deformation 的后置定语。

Glossary of Terms

1. blanking die　冲裁模
2. blanking clearance, die clearance　冲裁间隙
3. blanking force　冲裁力
4. piercing die　冲孔模
5. die, stamping and punching die　冲模
6. die life　冲模寿命
7. die shut height　模具闭合高度
8. tonnage of press　压力机吨位
9. shut height of press machine　压力机闭合高度
10. clearance between punch and die　凸凹模间隙
11. tolerance of fit　配合公差
12. shearing force diagram　剪力图
13. peak die load　模具最大负荷
14. outer slide　外滑块
15. center of die, center of load　压力中心
16. clamping force (element, device, piston)　夹紧力 (件，装置，活塞)
17. clamp plate (ring)　压板 (夹紧环)
18. shearing force (plane)　剪切力 (平面)
19. side edge　侧刃
20. side core　侧型芯
21. side clearance angle　侧隙角
22. side locating face　侧定位面
23. side-push plate　侧压板
24. shuttle table　移动工作台
25. matrix plate　凹模固定板
26. material removal rate　材料切除率
27. sheet metal　板料
28. sheet forming　板料成形，冲压

Reading Materials

Cutting Forces

The force required to cut (shear) the work material can be calculated by using the following formulas:

$$P = SLT \qquad \text{or} \qquad TN = \frac{SLT}{2000} \qquad \text{(for contours)}$$

$$P = \pi DST \qquad \text{or} \qquad TN = \frac{\pi DST}{2000} \qquad \text{(for round holes)}$$

Stripping Forces

The force required to strip the work material off punches can be calculated by using the following formulas:

$$P_s = 3500\,LT \qquad TN_s = \frac{3500LT}{2000} \qquad \text{(for contours)}$$

$$P_s = 3500\pi DT \quad TN_s = \frac{3500\pi DT}{2000} \qquad \text{(for round holes)}$$

Where P—Cutting force in pounds;

    TN—Cutting force in tons;

    P_s—Stripping force in pounds;

    TN_s—Stripping force in tons;

    S—Shear strength in pounds per square inch;

    L—Length of cut in inches;

    T—Thickness of material in inches;

    D—Diameter in inches.

Press Tonnage

This is a total of forces required to cut and form the part with a 30% safety factor added. In many cases, you will have to add stripping force if stripping is being done with a spring-loaded stripper, because the press has to compress the springs while cutting the material. Likewise, any spring pressure for forming, draw pads, and the like, will have to be added.

The force needed to punch a 2" (50.8mm) diameter hole in 1/8" (3.18mm) thick SAE1020 steel having a shear strength of 60,000 psi (413.7N/mm^2).

Lesson 11   Piercing and Blanking Die Design

Text

Piercing Die Design. A complete press tool for cutting two holes in work material at one *stroke* of the press, as classified and standardized by a large manufacturer as a single-station piercing die is shown in Figure 11-1.

Any complete press tool, consisting of a pair (or a combination of pairs) of mating members for producing pressworked (stamped) parts, including all supporting and *actuating* elements of the tool, is a die. Pressworking *terminology* commonly defines the female part of any complete press tool as a die.

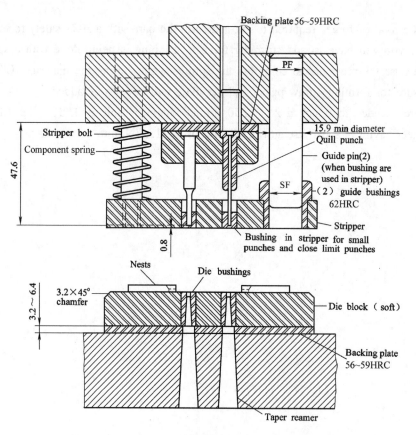

Figure 11-1   Typical single-station die for piercing holes

The guide *pins*, or *posts*, are mounted in the lower shoe. The upper shoe contains *bushings* which slide on the guide pins. The *assembly* of the lower and upper shoes with guide pins and bushings is a die set. Die sets in many sizes and designs are *commercially available*. The guide pins shown in Figures 11-2 and 11-3 guide the stripper in its vertical travel. For *clarity*, the guide pins are not shown in Figure 11-3.

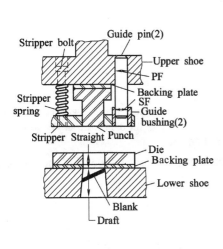

Figure 11-2　A simple blanking die

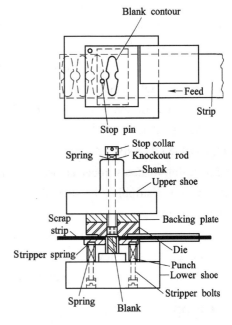

Figure 11-3　An inverted blanking die

A punch holder mounted to the upper shoe holds two round punches (male members of the die) which are guided by bushings *inserted* in the stripper[1]. A *sleeve*, or quill, encloses one punch to prevent its buckling under pressure from the ram of the press. After penetration of the work material, the two punches enter the die bushings for a slight distance.

The female member, or die, consists of two die bushings inserted in the die block. Since this press tool punches holes to the diameters required, the diameters of the die bushings are larger than those of the punches by the amount of clearance.

Since the work material stock or workpiece can *cling* to a punch on the upstroke, it may be necessary to strip the material from the punch. Spring-loaded strippers hold the work material against the die block until the punches are withdrawn from the punched holes. A workpiece to be pierced is commonly held and located in a nest (Figure 11-1) composed of flat plates shaped to *encircle* the outside part contours. Stock is positioned in dies by pins, blocks, or other types of stops for locating before the downstroke of the ram.

Blanking Die Design. The design of a small blanking die shown in Figure 11-2 is the same as that of the piercing die of Figure 11-1 except that a die replaces the die bushings and the two piercing punches are replaced by one blanking punch. A stock stop is *incorporated* instead of *nest* plates. This is a drop-through design since the finished blanks drop through the die, the lower shoe, and the press bolster.

Large blanks are commonly produced by an *inverted* blanking die (Figure 11-3) in which the die is mounted to the upper shoe with the punch *secured* to the bottom shoe. The passing of a large blank through the ***bolster*** is often impractical but its size may ***necessitate*** sectional die design (Figure 11-4).

Draft, or angular clearance in an inverted die is unnecessary because the blank does not pass through it. For ease of construction, regrinding, and strength the cutting edges of each section should not include points and ***intricate*** contours. Sections 1 and 2 of the die of Figure 11-4 were laid out to include the entire semicircular contour, with straight contours included in the other six sections.

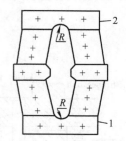

Figure 11-4   Eight-section layout
for die shown in Figure 11-3

The spring-loaded stripper is mounted on the lower shoe; it travels upward in stripping the stock from the punch fastened to the lower shoe. Stripper bolts hold and guide the stripper in its travel.

On the upstroke of the ram, the upper end of the knockout ***rod*** strikes an arm on the press frame, which forces the lower end of the rod downward, through the die, and ejects the finished blank from the die cavity. A stop ***collar*** retains the rods and limits their travel.

Questions

1. What is a complete single-station piercing die?
2. What is a die set?
3. What are the differences between piercing die and blanking die?
4. In the design of a large blanking die, why is angular clearance not required?

New Words and Expressions

1. stroke/strəuk/*n.* 打，击，敲，冲程，行程；*vt.* 划短线于，删掉

2. actuate/'æktjueit/*vt.* 开动，激励，驱使

3. terminology/ˌtəːmi'nɔlədʒi/*n.* 术语，专门名词

4. pin/pin/ *n.* 针，钉，栓，销

5. post/pəust/*n.* 柱，杆，桩；*vt.* 告示，揭示

6. bush/buʃ/*n.* 衬套，轴瓦；*vt.* 加衬套于……上

7. assembly/ə'sembli/*n.* 集合，集会，装配，部件

8. commercial/kə'məːʃəl/ *a.* 工业用的，工厂的，大批生产的，商业的

9. available/ə'veiləbl/*a.* 可用的，合用的，通用的，有效的

10. clarity/'klæriti/*n.* 透明

11. insert/in'səːt / *vt.* 插入，嵌入，投入；*n.* 金属型芯，衬垫

12. sleeve/sliːv/*n.* 袖套，套管，套筒；给……装套筒

13. cling/kliŋ/*vi.* 黏住，缠住，依附，靠拢

14. encircle/in'səːkl/*vt.* 包围，环绕，绕……旋转

15. incorporate/in'kɔːpəreit/*vt.* 结合，合并，编入，加入，使混合

16. nest/nest/*n.* 定位孔，窝，巢

17. invert/in'vəːt/*vt.* 倒转，倒装，倒置，转换；*n.* 倒转物，倒置物

18. secure/siˈkjuə/a. 安全的，牢固的，可靠的；vt. 保证，使安全，防护

19. bolster/ˈbəulstə/n. 长枕，垫枕，垫木；vt. 支撑，垫

20. necessitate/niˈsesiteit/vt. 需要，使成为必要，以……为条件

21. draft/drɑːft/n. 草稿，草图，斜度，牵引，通风；vt. 起草，设计

22. intricate/ˈintrikit/a. 错综的，复杂的，难懂的

23. rod/rɔd/n. 棒，杆，连杆，拉杆，推杆，测杆

24. collar/ˈkɔlə/n. 领，轭，环，垫圈，法兰盘

Notes

[1] A punch holder mounted to the upper shoe holds two round punches (male members of the die) which are guided by bushings inserted in the stripper.

安装在上模座上的凸模固定板固定两个圆形凸模（模具中的突出部分），这两个圆形凸模通过嵌入在卸料板上的模套进行导向。

句中分词短语 mounted to the upper shoe 作后置定语，它修饰的是主语 a punch holder，这里 mount 意为"安装"；另一个分词短语 inserted in the stripper 也是作后置定语，它修饰的是 bushings，insert 意为"嵌入"；由 which 引导出的定语从句修饰 two round punches。

Glossary of Terms

1. forming die   成形模
2. blanking die   冲裁模
3. assembling die   复合冲模，装配用模具
4. bed die   下模，底模
5. composite die, combined die   组合模，拼合模
6. compound die   复合模
7. compound blank and pierce dies   落料冲孔模
8. compound piercing die   复合冲孔模
9. shaving die   切边模，修边模
10. shankless die   无柄模具
11. scrapless progressive die   无废料连续模
12. return-blank type blanking die   顶出式落料模
13. restriking die   矫正模，校平模，整修模
14. reducing die   缩口模，缩径模
15. expanding die   胀形模，扩管模
16. cutting-off die   切断模
17. curling dies   卷曲模
18. die for special purpose   专用模
19. coining die   压印模
20. low-cost die   简易模
21. cavity plate（block）   凹模
22. local forming   局部成形
23. lancing die   切口模
24. sizing die   整形模
25. spinning   旋压
26. necking die   缩口或缩径模
27. flanging die   翻边模

Reading Materials

Stock Stops

In its simplest form, a stock stop may be a pin or small block against which an edge of the previously blanked opening is pushed after each stroke of the press. With sufficient clearance in the stock channel, the stock is momentarily lifted by its clinging to the punch, and is thus released from the stop. Figure 11-5 shows an adjustable type of solid block stop which can be moved along a support bar in increments up to 1" (25.4mm) to allow various stock lengths to be cut off.

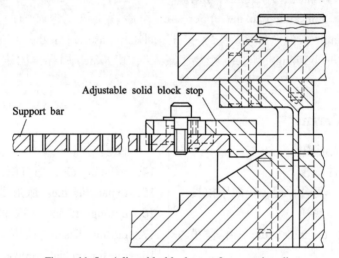

Figure 11-5   Adjustable block stop for a parting die

A starting stop, used to position stock as it is initially fed to a die, is shown in Figure 11-6,

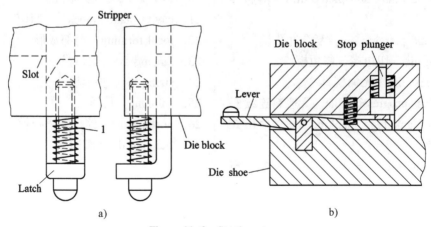

Figure 11-6   Starting stops

View a. Mounted on the stripper plate, the stop incorporates a latch which is pushed inward by the operator until its shoulder 1 contacts the stripper plate. The latch is held in to engage the edge of the incoming stock, the first die operation is completed, and the latch is released.

The starting stop shown at View b of Figure 11-6, mounted between the die shoe and die block, upwardly actuates a stop plunger to initially position the incoming stock. Compression springs return the manually operated lever after the first die operation is completed.

Trigger stops incorporate pivoted latches 1 (Figure 11-7, View a and View b). At the ram's descent, these latches are moved out of the blanked-out stock area by actuating pins 2. On the ascent of the ram, springs 3 control the lateral movement of the latch (equal to the side relief), which rides on the surface of the advancing stock and drops into the blanked area to rest against the cut edge of the cut-out area.

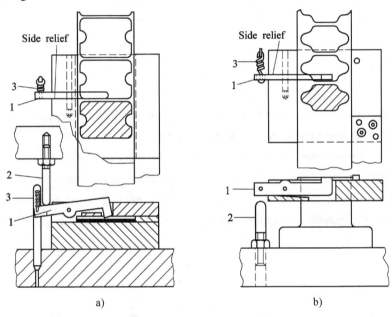

a)   b)

Figure 11-7   Trigger stops
a) Top stock engagement   b) Bottom stock engagement

Pilots

Since pilot breakage can result in the production of inaccurate parts and the jamming or breaking of die elements, pilots should be made of good tool-steel, heat-treated for maximum toughness and to a hardness of Rockwell C 57 to 60.

Press-fit Pilots. Press-fit pilots (Figure 11-8 and Figure 11-9, View c), which may drop out of the punch holder, are not recommended for high-speed dies but are often used in low-speed dies.

Pilots may be retained by methods shown in Figure 11-9. A threaded shank, shown at View a, is recommended for high-speed dies; thread length X and counterbore Y must be sufficient to allow for punch sharpening. For holes 3/4" (19mm) in diameter or larger, the pilot may be

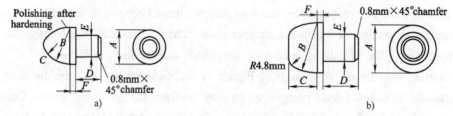

Figure 11-8　Press-fit pilots

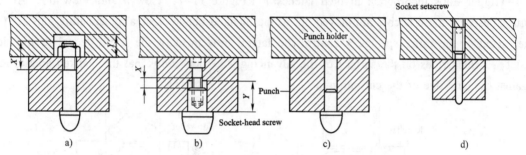

Figure 11-9　Methods of retaining pilots

a) Threaded shank　b) Screw-retained　c) Press-fit　d) Socket setscrew

held by a socket-head screw, shown at View b; recommended dimensions X and Y given for threaded-shank pilots also apply. A typical press-fit type is shown at View c. Pilots of less than 1/4" (6.4mm) diameter may be headed and secured by a socket setscrew, as shown at View d.

Indirect Pilots. Designs of pilots that enter holes in the scrap skeleton are shown in Figure 11-10. A headed design, at View a, is satisfactory for piloting in holes from 3/16" to 3/8" (4.8 ~ 9.5mm) in diameter. A quilled design, at View b, is suitable for pilots up to 3/16" (4.8mm) in diameter.

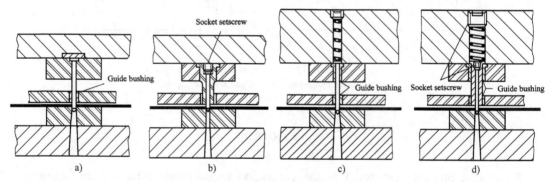

Figure 11-10　Indirect pilots

a) Headed　b) Quilled　c) Spring-backed　d) Spring-loaded quilled

Spring-loaded pilots should be used for stock exceeding No. 16 gage. A bushed, shouldered design is shown at View c of Figure 11-10. A slender pilot of drill rod shown at View d is locked in a bushed quill which is countersunk to fit the peened head of the pilot.

Tapered slug-clearance holes through the die and lower shoe should be provided, since indirect pilots generally pierce the strip during a misfeed.

Lesson 12   Compound and Progressive Die Design

Text

Compound Die Design. A compound die performs only cutting operations (usually blanking and piercing) which are completed during a single press stroke. A compound die can produce pierced blanks to close flatness and dimensional tolerances. A characteristic of compound dies is the inverted position of the blanking punch and blanking die. As shown in Figure 12-1, the die is fastened to the upper shoe and the blanking punch is mounted on the lower shoe. The blanking punch also functions as the piercing die, having a tapered hole in it and in the lower shoe for *slug disposal*.

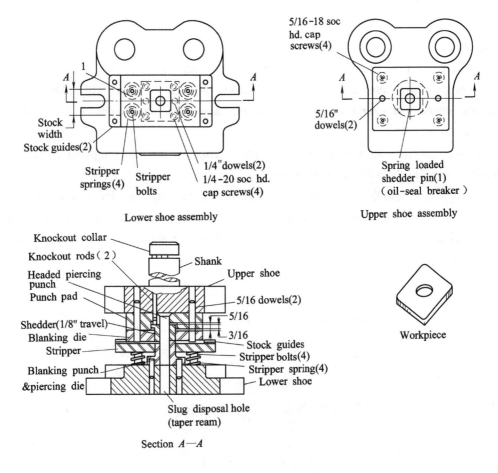

Figure 12-1   A compound die

On the upstroke of the press slide, the knockout bar of the press strikes the knockout collar, forcing the knockout rods and **shedder** downward, thus pushing the finished workpiece out of the blanking die[1]. The stock strip is guided by stock guides screwed to the spring-loaded stripper. On the upstroke the stock is stripped from the blanking die by the upward travel of the stripper. Before the cutting cycle starts, the strip stock is held flat between the stripper and the bottom surface of the blanking die.

Four special shoulder screws (stripper bolts), commercially available, guide the stripper in its travel and retain it against the preload of its springs.

The blanking die as well as the punch pad is screwed and *doweled* to the upper shoe.

A spring-loaded shedder pin (oil-seal breaker) incorporated in the shedder is *depressed* when the shedder pushes the blanked part from the die. On this upstroke of the ram the shedder pin breaks the oil seal between the surfaces of the blanked part and shedder, allowing the part to fall out of the blanking (upper) die.

Progressive Die Design. A progressive die performs a series of *fundamental* sheet metal operations at two or more stations during each press stroke in order to develop a workpiece as the strip stock moves through the die[2]. This type of die is sometimes called cut-and-carry, follow, or *gang* die. Each working station performs one or more *distinct* die operations, but the strip must move from the first through each succeeding station to produce a complete part. One or more *idle* stations may be *incorporated* in the die, not to perform work on the metal, but to locate the strip, to *facilitate* interstation *strip* travel, to provide maximum-size die sections or to simplify their construction.

The linear travel of the strip *stock* at each press stroke is called the progression, advance, or *pitch* and is equal to the interstation distance.

Parts requiring multiple operations to produce can be made at high production rates by progressive dies. The sheet metal is fed through as a coil strip, and different operation (such as punching, blanking, and notching) is performed at the same station of the machine with each stroke of a series of punches (Figure 12-2).

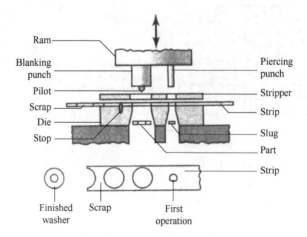

Figure 12-2　Schematic illustration of making a washer in a progressive die

Questions

1. What is a compound die?
2. What are the major characteristics of compound dies?
3. What is a progressive die?
4. What are the major characteristics of progressive dies?
5. What operations can be performed in a progressive die?

New Words and Expressions

1. slug /slʌg/ *n.* 弹丸，子弹，金属小块，废料，冷料
2. disposal /dis'pəuzəl/ *n.* 配置，排列，安排；处理，处置
3. shedder /'ʃedə/ *n.* 卸件装置，推（或拨、抛）料机
4. dowel /'dauəl/ *n.* 木钉，暗销；安装销钉，定位桩；*vt.* 用暗销接合
5. depressed /di'prest/ *a.* 降低的，凹陷的
6. fundamental /fʌndə'ment(ə)l/ *a.* 基本的，主权的；*n.* 基础，基本原理
7. gang /gæŋ/ *n.* 一组，一套
8. distinct /dis'tiŋkt/ *a.* 个别的，性质不同的；清楚的，显著的
9. idle /'aidl/ *a.* 闲的，无用的，空转的；*vt.* 使空转
10. incorporate /in'kɔːpəreit/ *vt.* 结合，合并，加入；*a.* 合并的，混合的
11. facilitate /fə'siliteit/ *vt.* 使便利，简化，推动，促进
12. strip /strip/ *vt.* 脱，拆卸，折断；*n.* 条，带，细长片
13. stock /stɔk/ *n.* 托盘，托柄，原料，备料；*vt.* 给……装柄
14. pitch /pitʃ/ *n.* 投掷，倾斜，螺距，节距
15. fastened to 固定到……上
16. be mounted on 安装在……上
17. a series of 一系列，许多
18. incorporated with 合并，混合
19. as distinct from 与……不同

Notes

[1] On the upstroke of the press slide, the knockout bar of the press strikes the knockout collar, forcing the knockout rods and shedder downward, thus pushing the finished workpiece out of the blanking die.

在压力机滑块的上行过程中，压力机的打杆碰到打料环，作用在打杆上的力使卸料装置下移，结果将成品件从落料凹模中推出。

句中 thus pushing the finished workpiece out of the blanking die 为 "thus +. ing" 形式，表示的是结果状语。可译为 "结果将成品件从落料凹模中推出"。

[2] A progressive die performs a series of fundamental sheet metal operations at two or more stations during each press stroke in order to develop a workpiece as the strip stock moves through the die.

级进模是指在压力机的每一次行程中，在模具的两个或更多的工位上同时完成一系列基本的金属板料冲压工序操作，其目的是使金属板料在通过模具后生产出完整的工件。

句中 in order to develop a workpiece as the strip stock moves through the die 为 "in order + 不定式" 形式，表示目的状语。可译为 "其目的是使金属板料在通过模具后生产出完整的工件"。

Glossary of Terms

1. insert die　镶拼模
2. trimming die　整修模
3. gang die　复式模，冲模组
4. supporting die　支承模
5. single（simple）operation die　单工序模（简单模）
6. progressive（continuous）die　级进模（连续模）
7. compound die　复合模
8. single-action die　单动（点）模
9. double- action die　双动（点）模
10. single-station piercing die　单工位冲孔模
11. inverted blanking die　倒置落料模
12. punch holder　凸模座
13. die holder　凹模座
14. lower shoe（bolster）下模座
15. upper shoe（bolster）上模座
16. inner ram　内滑块
17. outer ram　外滑块
18. knockout collar　打料环
19. knockout bar　打杆
20. spring-loaded pad　弹性垫板
21. shedder pin　卸料销
22. guide（stop）pin　导向（挡料）销
23. blank holder　压边装置
24. kicker　抛掷器，喷射器
25. stock stop　挡料板
26. press stroke　压力机行程

Reading Materials

Parts and Components of the Die

According to the preceding analysis on the types of the die structure, the die parts can be classified into two categories by its function. Such as technological structure parts and assistant structure parts. The classification of the die is shown in Table 12-1.

Table 12-1　Classification of the die

Number	Classification		Name of parts and components
1	Technological structure parts	Working parts	Punch, die, punch-die
		Locating parts	Stop pin and pilot, stock guide（guide rule）, locating pin（locating plate）, side guide, kicker
		Stripping, holding and knocking out parts	Stripper, blank holder, ejector, knockout rod

Number	Classification		Name of parts and components
2	Assistant structure parts	Guiding parts	Pillar, guide bushing, guide plate, guide tube
		Supporting and clamping parts	Upper and lower bolster, fastening plate of the punch and die, shank, bolster plate, limiter
		Fastening and other parts	Bolt, pin, others

Notes: 1. These parts take part in the performance of the technological process and contact with the blank directly.

2. These parts neither take part in the technological process nor contact with the blank directly.

Methods of Punch Support

Figure 12-3 presents a number of methods to support punches to meet various production requirements:

View a. When cutting punch A is sharpened, the same amount is ground off spacer B to maintain the relative distance C.

View b. If delicate punches must be grouped closely together, a hardened guide block with the required number of holes should be used.

View c. A slender piercing punch (at right) should be made shorter than an adjacent large punch.

View d. If punches must protrude more than 4" (102mm) beyond the punch holder, an auxiliary plate may be used to maintain stiffness.

View e. Flange width A of the punch should be greater than height B to provide stability for unguided punches.

View f. In a large punch, push-off pins can prevent slugs from pulling up and causing trouble.

View g. To avoid cracking a large, hardened punch or a punch plate, do not press a small punch directly into either of these members, instead, use a soft plug or insert.

View h. Long slotting punches should be hollow ground so that dimension A equals the metal thickness, so as to put shear on the punch. The ends should be flat for 1/8" (3.2mm) to avoid bending the stock.

View i. A quill is useful for supporting pin punches.

View j. A bushing in the stripper plate can guide the quill for increased punch support.

View k. Quills need not be limited to a single punch. If prevented from turning, they can be used for pin punches on close centers.

View l. Two quills are used for a bit punch, one to support the punch, the other to support the inner quill, when a stripper is not used.

View m. A dowel can be used to prevent rotation of the punches.

View n. For high-speed dies, a flat on the punch head is more positive.

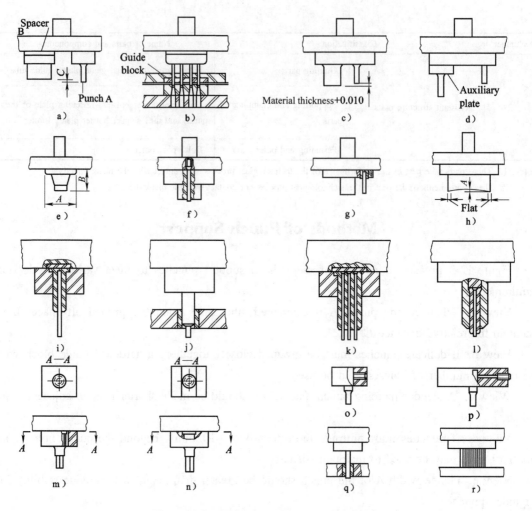

Figure 12-3   Various methods of punch support

View o. In low-production dies, a setscrew is adequate to hold the punch.

View p. When a deep hole must be drilled, a drill-rod pin can be used to span the distance.

View q. Light drill-rod punches are guided in the stripper plate to prevent buckling.

View r. Several punches can be set at close center distances.

Lesson 13   Strippers and Knockouts

Text

Strippers.   There are two types of strippers: fixed or spring-operated. The primary function of either type is to strip the workpiece from a cutting or noncutting punch or die. A stripper that forces a part out of a die may also be called a knockout, an inside stripper, or an ejector. Besides its primary function, a stripper may also hold down or ***clamp***, position, or guide the sheet, strip, or workpiece[1].

The stripper is usually the same width and length as the die block. In simpler dies, the stripper may be fastened with the same screws and dowels that fasten the die block, and the screwheads will be ***counterbored*** into the stripper. In more complex tools and with sectional die blocks, the die block screws will usually be ***inverted***, and the stripper fastener will be independent.

The stripper thickness must be sufficient to withstand the force required to strip the stock from the punch, plus whatever is required for the stock strip channel. Except for very heavy tools or large blank areas, the thickness required for screwhead counterbores, in the range of 3/8" to 5/8" (9.5 ~ 16mm), will be sufficient.

The height of the stock strip channel should be at least 1.5 times the stock thickness. This height should be increased if the stock is to be lifted over a fixed pin stop. The channel width should be the width of the stock strip, plus adequate clearance to allow for variations in the width of the strip cut.

Choice of the methods of applying springs to stripper plates depends on the required pressure, space ***limitations***, shape of the die, nature of the work, and production requirements. Figure 13-1 presents a number of such methods.

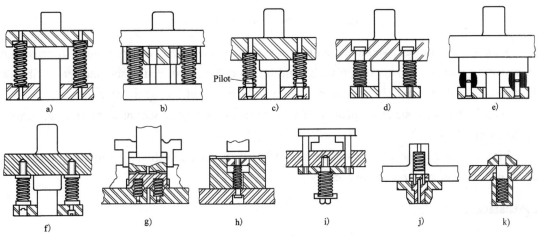

Figure 13-1   Use of die springs with strippers

Knockouts.   Since the cut blank will be retained in the die block by friction, some means of ejecting on the ram **upstroke** must be provided. A knockout assembly consists of a plate, a push rod, and a retaining **collar**. The plate is a loose fit with the die opening contour, and moves upward as the blank is cut. Attached to the plate, usually by rivets, is a heavy push rod which slides in a hole in the **shank** of the die set. This rod projects above the shank, and a collar retains and limits the **stroke** of the assembly. Near the upper limit of the ram stroke, a knockout bar in the press will contact the push rod and eject the blank.

It is essential that the means of retaining the knockout assembly be **secure**, since serious damage would otherwise occur[2].

In the ejection of parts, positive knockouts offer the following advantages over spring strippers where the part shape and the die selections allow their use:

1. **Automatic part disposal.** The blank, ejected near the top of the ram stroke can be blown to the back of the press, or the press may be inclined and the same result obtained.

2. **Lower die cost.** Knockouts are generally of lower cost than spring strippers.

3. **Positive action.** Knockouts do not stick as spring strippers occasionally do.

4. **Lower pressure requirements.** There are no heavy springs to be compressed during the ram descent.

Figure 13-2 shows several good knockout designs.

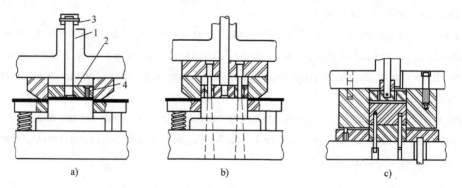

Figure 13-2   Positive knockouts for dies

View a. This design, applied to a plain inverted compound die, is very simple. It consists of an **actuating plunger** 1, knockout plate 2, and a stop collar 3 doweled to the plunger 1. Shedder 4 consists of a shouldered pin backed by a spring which is confined by a setscrew.

View b shows the knockout plate used as a means of guiding slender piercing punches through hardened bushings.

View c shows a design in which the **flanged** shell, upon completion, is carried upward in the upper die and ejected by a positive knockout.

Questions

1. What are two types of strippers?

2. What are the functions of stripper?

3. Describe choice of the methods of applying springs to stripper plates.

4. What is a knockout?

5. What are the functions of knockout?

6. What are the four advantages of positive knockouts over spring strippers?

New Words and Expressions

1. stripper /ˈstripə/ *n.* 卸料板，冲孔模板

2. clamp /klæmp/ *n.* 夹具，夹钳；压板，压铁；*vt.* 夹紧，钳住

3. counterbore /ˈkauntəˈbɔː/ *n.* 平底钻，沉孔，埋头孔；*v.* 钻平底孔

4. invert /inˈvəːt/ *vt.* 倒转，倒置，反向；*n.* 倒置物，转向

5. limitation /limiˈteiʃən/ *n.* 限制，局限性；极限，缺点

6. upstroke /ˈʌpˈstrəuk/ *n.* 向上的一击，上升冲程

7. collar /ˈkɔlə/ *n.* 环，轴环，套环；垫圈

8. shank /ʃæŋk/ *n.* 柄，杆

9. stroke /strəuk/ *n.* 打，击，敲；行程

10. secure /siˈkjuə/ *a.* 安全的，可靠的；

vt. 保证，使安全，防护

11. actuate/ˈæktʃueit/ *vt.* 开动，激励，驱使

12. plunger /ˈplʌndʒə/ *n.* 柱塞，活塞

13. flange /flændʒ/ *n.* 法兰，边缘，凸缘；*vt.* 装凸缘，镶边

14. fasten on 握住，抓牢；把……钉在……上

15. be essential to 对……很重要

16. be sufficient to 足以

17. depend on (upon) 依靠，取决于

18. be attached to 附属于

19. as a means of 作为……的工具

20. be actuated by ……为……所驱使

21. actuating motor 伺服电动机

Notes

[1] Besides its primary function, a stripper may also hold down or clamp, position, or guide the sheet, strip, or workpiece.

除基本功能外，卸料板还具有压料即夹紧，定位，导板料、条料和工件的作用。

句中 besides 为介词，可译为"除……以外，还……"；besides its primary function 为介词短语，可译为"除（它的）基本功能外"。

[2] It is essential that the means of retaining the knockout assembly be secure, since serious damage would otherwise occur.

因为一些危险在偶然情况下可能会发生，因此，保持打料装置的可靠性是很重要的。

句中 It is essential that...，it 作为从句的先行代词，可译为"……是很重要的"；在 since serious damage would otherwise occur 中，since 引导原因状语从句，可译为"因为在相反情况下可能会发生一些危险"。

Glossary of Terms

1. working (locating) part　工作（定位）零件
2. fastening (guiding) part　固定（导向）零件
3. supporting part　支承零件
4. guide plate　导板
5. guide pillar (pin)　导柱
6. guide tube (bushing)　导筒（套）
7. fixed stripper　固定卸料板
8. elastic (spring) stripper　弹性（弹簧）卸料板
9. fixed stop pin　固定挡料销
10. moved stop pin　活动挡料销
11. pilot (guide) pin　导正销，定位销
12. ejecting pin　顶杆
13. stock (side) guide　导料（侧压）板
14. bolster (ejecting) plate　垫（顶）板
15. strip (sheet) stock　条（片）料
16. strip feeder　送料装置
17. punch pad　凸模垫板（上垫板）
18. die pad　凹模垫板（下垫板）
19. stamping press　冲压机
20. hold down　压制，抑制
21. automatic feeder　自动送料装置
22. technological property　工艺性能
23. technological design　工艺设计
24. dimension marking　尺寸标注
25. locating datum　定位基准
26. roundness radius　圆角半径
27. less waste arrangement　无废料排样
28. scrap size　排样搭边

Reading Materials

Methods of Reducing Cutting Forces

Since cutting operations are characterized by very high forces exerted over very short periods of time, it is sometimes desirable to reduce the force and spread it over a longer portion of the ram stroke. Punch contours of large perimeter or many smaller punches will frequently result in tonnage requirements beyond the capacity of an available press. Also, whenever abnormally high tonnage requirements are concentrated in a small area, design difficulties are increased.

Two methods are generally used to reduce cutting forces and to smooth the shock impact of heavy loads. Keep in mind that during a piercing operation with proper clearance, complete fracture occurs when the punch has penetrated one-third of the material thickness.

1. By stepping punches one-third of the material thickness, they can cut individually. Typically using three punches of the same diameter stepped properly, you use one-third of the total tonnage required to do all three simultaneously.

2. Adding shear to the die or punch equal to one-third of the material thickness reduces the tonnage required by 50% for that area being cut with shear applied. Note that shear is applied to the die member (punch or die) that contacts the scrap. Therefore, deformation due to the shear

angles does not affect the part (Figure 13-3).

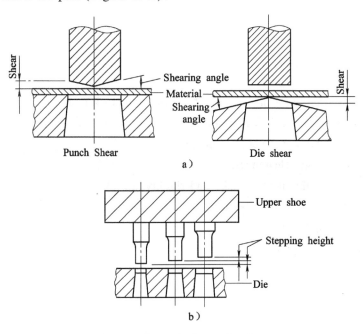

Figure 13-3   Reducing cutting forces

In piercing, the direction of the shear angles must be such that the cut proceeds from the outer extremities of the contour toward the center. This avoids stretching the material before it is cut free.

Note: Using tunnel strippers whenever possible, instead of spring-loaded strippers, will also keep the press load to a minimum.

Punch Dimensioning

The determination of punch dimensions has been generally based on practical experience.

When the diameter of a pierced round hole equals stock thickness, the unit compressive stress on the punch is four times the unit shear stress on the cut area of the stock, from the following formula:

$$\frac{4S_s t}{S_c d} = 1$$

Where S_c—unit compressive stress on the punch, psi;

S_s—unit shear stress on the stock, psi;

t—stock thickness, inch;

d—diameter of punches hole, inch.

The diameters of most holes are greater than stock thickness; a value for the ratio d/t of 1. 1 is recommended.

The maximum allowable length of a punch can be calculated from the formula:

$$L = \frac{\pi dEd^{1/2}}{8S_s t}$$

Where d/t—1. 1 or higher;

E—modulus of elasticity.

This is not to say that holes having diameters less than stock thickness cannot be successfully punched. The punching of such holes can be facilitated by many methods:

1. Punch steels of high compressive strengths.
2. Greater than average clearances.
3. Optimum punch alignment, finish, and rigidity.
4. Shear on punches or dies or both.
5. Prevention of stock slippage.
6. Optimum stripper design.

Lesson 14　Bending Dies

Text

Bending is the uniform ***straining*** of material, usually flat sheet or strip metal, around a straight axis which ***lies*** in the ***neutral*** plane and normal to the ***lengthwise*** direction of the sheet or strip. Metal flow takes place within the plastic range of the metal, so that the bend retains a permanent set after removal of the applied stress[1]. The ***inner*** surface of a bend is in compression; the outer surface is in tension. A pure bending action does not reproduce the exact shape of the punch and die in the metal; such a reproduction is one of forming. The neutral axis is the plane area in bent metal where all strains are zero.

Bend Radii. Minimum bend radii vary for different metals; generally, different annealed metals can be bent to a radius equal to the thickness of the metal without ***cracking*** or weakening.

Bend Allowances. Since bent metal is longer after bending, its increased length, generally of concern to the product designer, may also have to be considered by the die designer if the length tolerance of the bent part is critical. The length of bent metal may be calculated from the equation:

$$B = \frac{A}{360} 2\pi (R_i + Kt)$$

Where　B—bend allowance, inch (mm) (along neutral axis);

　　　　A—bend angle, deg;

　　　　R_i—inside radius of bend, inch (mm);

　　　　t—metal thickness, inch (mm);

　　　　K—0.33 when R_i is less than $2t$ and is 0.50 when R_i is more than $2t$.

Bending Methods. Two bending methods are commonly made use of in press tools. Metal sheet or strip, supported by a V block (Figure 14-1a), is forced by a ***wedge***-shaped punch into the block. This method, termed V bending, produces a bend having an included angle which may be ***acute***, ***obtuse***, or of 90°. ***Friction*** between a spring-loaded ***knurled*** pin in the ***vee*** of a die and the part will prevent or reduce side ***creep*** of the part during its bending[2].

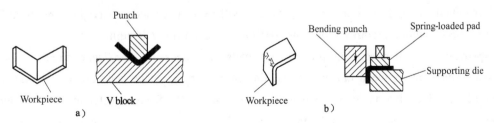

Figure 14-1　Bending methods
a) V bending　b) Edge bending

Edge bending (Figure 14-1b) is ***cantilever*** loading of a beam. The bending punch 1 forces the metal against the supporting die 3. The bend axis is parallel to the edge of the die. The workpiece is ***clamped*** to the die block by a spring-loaded pad 2 before the punch ***contacts*** the workpiece to prevent its movement during downward travel of the punch. V-bending die and edge-bending die are shown in Figure 14-2.

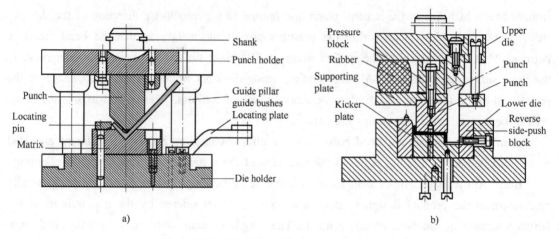

Figure 14-2   V-bending die and edge-bending die

a) V-bending die   b) Edge-bending die

Bending Force.   The force required for V bending is as follows:

$$P = \frac{KLSt^2}{W}$$

Where P —bending force, tons (for ***metric*** usage, multiply number of tons by 8. 896 to obtain kilonewtons);

$\quad\quad K$ —die opening factor: 1. 20 for a die opening of 16 times metal thickness, 1. 33 for an opening of eight times metal thickness;

$\quad\quad L$ —length of part, inch;

$\quad\quad S$ —ultimate tensile strength, tons per square inch;

$\quad\quad W$ —width of V or U die, inch;

$\quad\quad t$ —metal thickness, inch.

For U bending (channel bending) pressures will be approximately twice those required for V bending; edge bending requires about one-half those needed for V bending.

Springback.   After bending pressure on metal is ***released***, the elastic stresses also are released, which causes metal movement resulting in a decrease in the bend angle (as well as an increase in the included angle between the bent ***portions***)[3]. Such a metal movement, termed springback, varies in steel from 0. 5° to 5°, depending upon its hardness, phosphor bronze may spring back from 10° to 15°.

V-bending dies ***customarily*** compensate for springback with V blocks and wedge-shaped punches having included angles somewhat less than that required in the part. The part is bent through a greater angle than that required but it springs back to the desired angle.

Parts produced in other types of bending dies are also overbent through an angle equal to the spring-back angle with an undercut or *relieved* punch.

Questions

1. What is the basic principle involved in a bending die?
2. What causes springback?
3. What is minimum bend radii?
4. Describe the difference between V bending and edge bending.
5. What are the influence factors of bending force?

New Words and Expressions

1. strain/strein/*vt.* 拉紧，伸张；*vi.* 弯曲，变形，拉紧；*n.* 拉紧，张力，应变

2. lie/lai/*vi.* 躺，保持……状态，处在，位于；*n.* 位置，状态

3. neutral/'nju:trəl/*a.* 中立的，中性的；*n.* 中线，空挡

4. lengthwise/'leŋθwaiz/*ad.* 纵长地；*a.* 纵长的

5. inner/ 'inə /*a.* 内部的，内心的；*n.* 内部，里面

6. crack/kræk/*vt.* 弄裂，敲碎；*vi.* 破裂，断裂；*n.* 裂缝；*a.* 第一流的，精炼的

7. wedge/wedʒ/ *n.* 劈，楔块，楔形物，浇口；*v.* 楔入

8. acute /ə'kju:t/ *a.* 锐角的，敏锐的，急剧的

9. obtuse/əb'tju:s/ *a.* 钝（角）的，迟钝的，愚钝的

10. friction/'frikʃən/*n.* 摩擦（力），阻力

11. knurl /nə:l/*n.* 节，压纹，滚花；*vt.* 滚花，压花

12. vee/vi:/ *n.* V 字形物；*a.* V 字形的

13. creep /kri:p/*vi.* 爬，蔓延；*n.* 蠕变，位移

14. cantilever /'kæntili:və/ *n.* 悬臂，交叉支架，角撑架

15. clamp /klæmp/ *n.* 夹钳，夹具，压板，压铁；*vt.* 夹紧，钳住

16. contact /'kɔntækt/*n.* 接触，联系，触点；*v.* 接触，联系；*a.* 保持接触的

17. metric/'metrik/*a.* 公制的，米制的

18. release/ri'li:s/*vt.* 释放，解放，放松，放出；*n.* 释放装置，排气装置

19. portion/'pɔ:ʃən/*n.* 部分，一份；*vt.* 分配

20. customarily/'kʌstəmərili/*ad.* 照例，通常，习惯上

21. relieve/ri'li:v/ *vt.* 援救，减轻，缓和，解除

22. flat sheet or strip metal 平钢板或金属带材

23. lie in 在于

24. neutral plane 中性面

25. take place 发生

26. inner surface　内表面
27. outer surface　外表面
28. ultimate tensile strength　极限抗拉强度
29. elastic stress　弹性应力
30. spring-back angle　回弹角

Notes

[1] Metal flow takes place within the plastic range of the metal, so that the bend retains a permanent set after removal of the applied stress.

由于弯曲金属的流动发生在金属塑性变形范围内，因此当所施加的外力去除后将会保留一个永久性的弯曲变形。

句中的介词短语 within the plastic range of the metal 意为在金属的塑性变形范围内；so that 引导的是一个状语从句，它表示结果；单个分词 applied 作 stress 的定语，意为"外加的"。

[2] Friction between a spring-loaded knurled pin in the vee of a die and the part will prevent or reduce side creep of the part during its bending.

在 V 形模具内，弹性滚花压销和零件之间的摩擦将会阻碍或减小零件在弯曲过程中的边缘位移。

between 的一般用法为 between A and B，本句中 between 后面的两个关键词：一个是 pin，另一个是 part；在 pin 前有两个分词，一个是 spring-loaded，另一个是 knurled，二者都是 pin 的前置定语。

[3] After bending pressure on metal is released, the elastic stresses also are released, which causes metal movement resulting in a decrease in the bend angle (as well as an increase in the included angle between the bent portions).

当作用在金属板料上的弯曲力去除后，弹性应力也就随即消失，这将引起金属板料的移动，从而导致弯曲角减小（也即弯曲部件间的包角增大）。

句子中由 which 引导的从句是一个非限定性定语从句，它对整个主句 the elastic stresses also are released 作进一步说明，其中关系代词 which 在定语从句中作主语，由于 which 代替了整个主句，所以从句的谓语用 causes 这一形式。句中有一分词短语 resulting in a decrease in the bend angle 在定语从句中作状语，表示结果，可译为"从而导致弯曲角减小"。

Glossary of Terms

1. bend ability　可弯性
2. bend length　弯曲中性层弧长
3. bend arc　弯曲弧
4. bending angle（line）　弯曲角（线）
5. bending brake（bending machine）　弯板机，折弯机
6. bending fatigue　弯曲疲劳
7. bending fixture　弯曲夹具
8. bending force　弯曲力
9. bending radius　弯曲半径
10. minimum bending radius　最小弯曲半径
11. bending operation　弯曲工序
12. bending press　弯曲压力机，压弯机
13. bending strength　抗弯强度

14. bending test（stress，work）弯曲试验（应力，加工）
15. bent pilot 弯曲导正销
16. air-bend die 自由弯曲模
17. double V-die 双 V 形弯曲模
18. bending moment diagram 弯矩图
19. blank length of bend 弯曲件展开长度
20. relative bending radius 相对弯曲半径
21. bending equipment for plastics 塑料折弯设备
22. bending with sizing 校正弯曲
23. drawing machine 拉拔机
24. drawing numbers 拉深次数
25. drawing ratio（coefficient，force，speed）拉深比（系数，力，速度）
26. springback 回弹
27. superplastic forming 超塑性成形
28. explosive forming 爆炸成形
29. deep drawing 深拉成形

Reading Materials

Drawing Dies

Drawing is a process of changing a flat, precut metal blank into a hollow vessel without excessive wrinkling, thinning, or fracturing. The various forms produced may be cylindrical or box-shaped with straight or tapered sides or a combination of straight, tapered, or curved sides. The size of the parts may vary from 0.250" （6.35mm）diameter or smaller, to aircraft or automotive parts large enough to require the use of mechanical handling equipment.

Metal Flow. When a metal blank is drawn into a die, a change in its shape is brought about by forcing the metal to flow on a plane parallel to the die face, with the result that its thickness and surface area remain about the same as the blank. Figure 14-3 shows schematically the flow of metal in circular shells. The units within one pair of radial boundaries have been numbered and each unit moved progressively toward the center in three steps. If the shell were drawn in this manner, and a certain unit area examined after each depth shown, it would show （1）a size change only as the metal moves toward the die radius；（2）a shape change only as the metal moves over the die radius. Observe that no change takes place in area 1, and the maximum change is noted in area 5.

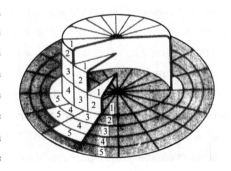

Figure 14-3   A step-by-step flow of metal

Single-Action Dies. The simplest type of draw die is one with only a punch and die. Each component may be designed in one piece without a shoe by incorporating features for attaching them to the ram and bolster plate of the press. Figure 14-4a shows a simple type of draw die in which the precut blank is placed in the recess on top of the die, and the punch descends, pushing the cup through the die. As the punch ascends, the cup is stripped from the punch by the counterbore in the bottom of the die. The top edge of the shell expands slightly to make this possible. The punch has an air vent to eliminate suction which would hold the cup on the punch and damage the cup when it is stripped from the punch.

The method by which the blank is held in position is important, because successful drawing is somewhat dependent upon the proper control of blank holder pressure. A simple form of drawing die with a rigid flat blank holder for use with 13-gage and heavier stock is shown in Figure 14-4b. When the punch comes in contact with the stock, it will be drawn into the die without allowing wrinkles to form.

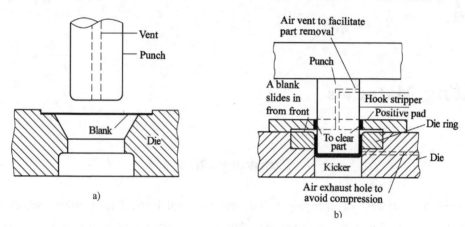

a)

b)

Figure 14-4　Draw die types

a) Simple type　b) Simple draw die for heavy stock

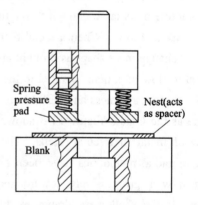

Figure 14-5　Draw die with spring pressure pad

Another type of drawing die for use in a single-action press is shown in Figure 14-5. This die is a plain single-action type where the punch pushes the metal blank into the die, using a spring-loaded pressure pad to control the metal flow. The cup either drops through the die or is stripped off the punch by the pressure pad. The sketch shows the pressure pad extending over the nest, which acts as a spacer and is ground to such a thickness that an even and proper pressure is exerted on the blank at all times. If the spring pressure pad is used without the spacer, the more the springs are depressed, the greater the pressure exerted on the blank, thereby limiting the depth of draw. Because of limited pressures obtainable, this type of die should be used with light-gage stock and shallow depths.

A single-action die drawing flanged parts, having a spring-loaded pressure pad and stripper, is shown in Figure 14-6. The stripper may also be used to form slight indentations or reentrant curves in the bottom of a cup, with or without a flange. Draw tools in which the pressure pad is attached to the punch are suitable only for shallow draws. The pressure cannot be easily adjusted, and the short springs tend to build up pressure too quickly for deep draws. This type of die is often constructed in an inverted position with the punch fastened to the lower portion of the die.

Deep draws may be made on single-action dies, where the pressure on the blank holder is more evenly controlled by a die cushion or pad attached to the bed of the press. The typical construction of such a die is shown in Figure 14-7. This is an inverted die with the punch on the die's lower portion.

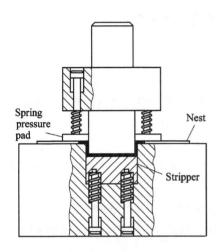

Figure 14-6 A draw die with spring
pressure pad and stripper

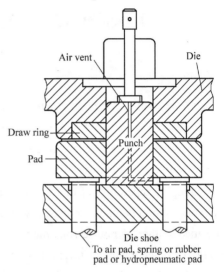

Figure 14-7   Cross section of an inverted
draw die for a single-action press

Notes: 1. Die is attached to the ram.

2. Punch and pressure pad are on the lower shoe.

Double-Action Dies. In dies designed for use in a double-action press, the blank holder is fastened to the outer ram which descends first and grips the blank; then the punch, which is fastened to the inner ram, descends, forming the part. These dies may be push-through type, or the parts may be ejected from the die with a knockout attached to the die cushion or by means of a delayed action kicker.

Unit Five

Lesson 15   Die-Casting Dies

Text

A pressure die casting die is an assembly of materials, mostly *ferrous* metals, each of which plays a part in a *mechanism* which will operate under conditions of rapidly changing temperatures, as the molten metal is *injected* under pressure and then immediately cooled[1]. Some parts of the die *merely* act as holding elements; others, such as the die inserts and the cores, have to withstand the *impact* and high temperature of the molten metal. The mechanisms for moving the *ejectors* and cores must work smoothly in the constantly changing temperature of a die. Parts which act as bearings are made either of non-ferrous metals such as phosphor bronze, or the steel surfaces are given a *nitride* or other treatment to resist wear.

The die insert for producing the shape of the casting has to be machined accurately and heat treated to provide the best possible mechanical properties at high temperature. Most die casting dies operate automatically and, while this leads to high productivity when there are no *delays*, the cost of breakdowns are *correspondingly* high. During the past *decade* the manufacture of dies has been *revolutionized* by *sophisticated* methods ranging from improved spark erosion to computer-controlled die making. Among other efforts the Science Research Council is supporting a large *scale* research programme to include die heat treatment and surface *coatings*, computer-aided die design, special machining processes and the economics of die manufacture.

A single die may contain 10 or more different steels plus several non-ferrous metals and special heat resisting alloys. Figures 15-1 and 15-2 illustrate schematically a die which typifies the various components, each of which involves design, engineering and metallurgical problems.

Low alloy steel and cast iron components. The ejector box is constructed of several *rectangular* blocks built up to contain the ejector plates which are guided by round section runner bars or guide *pillars*. The box sections, ejector plates and runner bars are of mild steel, with approximately 0.15% carbon. The guide pillars and ejector stops may be case hardened. The function of these parts is to support and guide, not to *endure* shock loading. Bolsters undergo mechanical impact and stress but not a great deal of thermal shock and are often made of medium carbon steel. *Alternatively* these parts are steel castings or they may be of a spheroidal graphite cast iron.

The die blocks contain the movement mechanisms of the die and the inserts incorporating the casting form so that the assembly can be mounted between the tie bars of the machine[2]. The blocks are usually made of medium carbon steel; a typical British specification is BS 970 08 M40 (En8), with 0.4% carbon, 0.8% manganese and 0.3% silicon. Sometimes a prehardened steel is used, a typical analysis being 0.35% carbon, 1.0% manganese, 0.5% silicon, 1.65% chromium and 0.5% molybdenum. This composition is covered by the American AISI specification P.20.

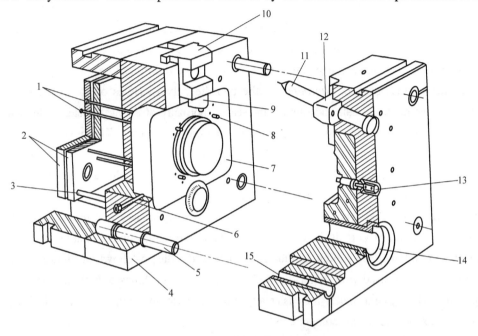

Figure 15-1   Typical die assembly

1—Ejector pins   2—Ejector plates   3—Ejector return pin   4—Base support   5—Guide pillar
6—Die insert waterway   7—Die insert   8—Fixed core   9—Moving core   10—Moving core holder
11—Angle pin   12—Core locking wedge   13—Cascade waterway   14—Plunger bush   15—Guide bush

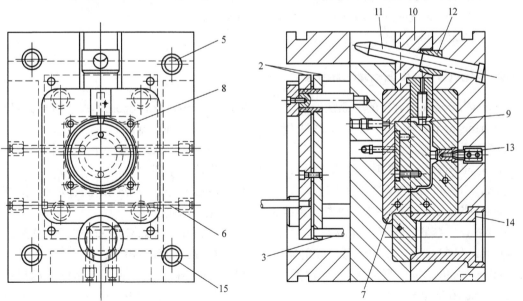

Figure 15-2   Views illustrating structural features of die shown in Figure 15-1

The die blocks are shaped to form the *recess* into which the inserts will be fitted. Alternatively, this is a procedure which is steadily gaining on the other method, large die blocks are supplied as medium carbon steel or cast iron, cast as closely as possible to the required shape. The blocks are ground flat so that the injected alloy will not escape under the considerable pressure of injection. It is preferable to have the parting line in one plane, but sometimes the design of a component makes it necessary to construct the die with irregular parting line surfaces to avoid creating an undercut. Once the parting line has been established and the blocks machined, guide pillars, usually between 10mm and 60mm in diameter, depending on the size of the die blocks, are incorporated into one half of the die to ensure exact *alignment* of the two die halves. Holes of an appropriate size are machined into the other die half and fitted with bushes, normally of carburized steel. On symmetrical dies, one pillar may be offset, to avoid assembling the die incorrectly. The pillars are normally of case hardened mild steel; occasionally case hardened nickel steels are used, for increased strength.

Where possible, guide pillars are fitted into the die half which holds any *protruding* die form (usually the moving, or ejector, half) to give protection to the cavity form when the die is taken from the machine for maintenance. If positioning in this way causes interference with a casting retrieval arm, particularly if it is of the fixed path extraction type, unable to offer lateral movement, one pillar is repositioned in the fixed half of the die, as shown in Figure 15-1. Square guide pillars are often used with larger dies to be operated on machines of 600 tonnes locking force and upwards. This system makes for easier adjustment, sometimes required on such dies, arising from thermal expansion of die components.

Questions

1. What is a pressure die casting die?
2. Of which materials are the die blocks usually made?
3. What is the pillar?

New Words and Expressions

1. ferrous/ˈferəs/ *a.* 含铁的，亚铁的
2. mechanism/ˈmekənizəm/ *n.* 机械装置，机构，机理，技巧
3. inject/inˈdʒekt/ *vt.* 注射，注入，引入
4. merely/ˈmiəli/ *ad.* 仅仅，只不过
5. impact/ˈimpækt/ *n.* 碰撞，冲击，影响，效果; *vt.* 装紧，压紧，冲击
6. ejector/iˈdʒektə/ *n.* 推出装置，喷射器
7. nitride/ˈnaitraid/ *n.* 氮化物; *v.* 渗氮
8. delay/diˈlei/ *n. & v.* 耽误，推迟

9. correspond/kɔriˈspɔnd/ *vi.* 相当，对应，符合（+to，with）
10. decade/ˈdekeid/ *n.* 十，十个一组，十进位，十年
11. revolutionize/revəˈluːʃənaiz/ *vt.* 使革命化，彻底改革
12. sophisticate/səˈfistikeit/ *vt.* 使复杂，使精致，曲解，掺杂
13. scale/skeil/ *n.* 天平，磅秤; 标尺，刻度，比例; 氧化铁皮; *vt.* 把……过

秤，测量，去锈

14. coating/ˈkəutiŋ/n. 涂层，覆盖层

15. rectangular/rekˈtæŋgjulə/a. 长方形的，矩形的，成直角的

16. pillar/ˈpilə/n. 柱，柱状物；vt. 用柱支撑，用柱加固

17. endure/inˈdjuə/vt. 耐，忍受，容忍；vi. 忍耐，持久

18. alternative/ɔ:lˈtə:nətiv/a. 可能的，交替的，两者挑一的；n. 替换物

19. recess/riˈses/n. 休息，凹槽，凹进处；vt. 使凹进；vi. 休息

20. alignment/əˈlainmənt/n. 排列，组合，校直，准线

21. protrude/prəˈtru:d/v. 伸出，突出，凸出

22. pressure die casting die 压铸模

23. play a part in 起作用

24. mechanical properties 机械性能

25. guide pillar 导柱

26. ejector plate 推板

27. spheroidal graphite cast iron 球墨铸铁

28. irregular parting line surface 不规则分型面

Notes

[1] A pressure die casting die is an assembly of materials, mostly ferrous metals, each of which plays a part in a mechanism which will operate under conditions of rapidly changing temperatures, as the molten metal is injected under pressure and then immediately cooled.

压铸模多数是由一些铁质金属材料部件组装而成的，且每一个部件都在机构中起一定的作用。熔融金属在一定压力下被注入模具型腔后立即冷却，致使这些部件在温度快速变化的条件下工作。

句中 mostly ferrous metals 是指 materials，第一个 which 也是指 materials，第二个 which 引导的定语从句修饰 mechanism。

[2] The die blocks contain the movement mechanisms of the die and the inserts incorporating the casting form so that the assembly can be mounted between the tie bars of the machine.

模板包括组成铸型型腔的凹模和型芯的运动机构，以便使模具能安装在压铸机的拉杆之间。

句中 so that 引出目的状语从句，译为"为了"或"以便"等。

Glossary of Terms

1. die-casting die 压力铸造模具（简称压铸模）

2. fixed clamping plate 定模座板

3. fix die, cover die 定模

4. moving die, ejector die 动模

5. moving clamping plate 动模座板

6. support plate, backing plate 支承板

7. movable core 活动型芯

8. fixed (moving) die insert 定（动）模镶块

9. sprue bush 浇口套

10. baffle 导流块

11. sprue spreader 分流锥（分流器）

12. ejector guide pin 推板导柱

13. ejector guide bush 推板导套

14. ejector pin (plate) 推杆（板）

15. ejector sleeve 推管
16. ejector pin retaining plate 推杆固定板
17. sprue 直浇道，主流道，浇口
18. runner 横浇道
19. gate 内浇口
20. overflow well 溢流槽
21. air vent 排气槽
22. parting line 分型面

23. feed（gating, runner）system 浇注系统
24. pouring temperature（rate, time）浇注温度（速度，时间）
25. sprue base（bush, gate, puller）直浇道窝（浇口套，直接浇口，拉料杆）
26. die block 凹模固定板

Reading Materials

The Die Insert

This is the heart of the die, since it contains the outside shape of the component, forming the cavity into which the molten metal will be injected. Die inserts, together with the cores which form recesses, must endure the effects of temperature and injection pressure of the molten metal. Each injection is a step towards the thermal fatigue which will eventually cause deterioration of the die. The cycle of stresses in the die insert results in a sudden increase of temperature as the molten metal enters, followed by a rapid decrease. Often the die insert is built up from several pieces, either to save materials when large blocks incorporate small projections, or to facilitate replacement of parts, or to ensure that an available heat treatment furnace will accommodate the die block. Spark erosion techniques, widely used nowadays to manufacture small or medium sized die components in one piece, avoid the danger of molten metal being forced down joins between separate parts.

Cast-to-form is now a well established procedure for making cast iron permanent mould dies, of sufficient accuracy for the requirements of that process. Die inserts for pressure die casting need to be made to an accuracy of the order of ±0.002mm. Furthermore, the inserts are usually to be made of 5% chromium steel and, so far, the difficulties of casting such a material to a sufficient accuracy has prevented "cast-to-form" from being used for pressure die casting dies.

Where deep inserts are required, clearances to assist fitting are provided and radii on the corners are included to improve strength and remove stress raisers, as indicated in Figure 15-3. For very deep inserts, filleting at two levels is advisable, both for ease of assembly and to assist heat transmission from the insert to the die block. The select of steel and its subsequent heat treatment are vital factors in ensuring optimum life and accuracy of the die casting die.

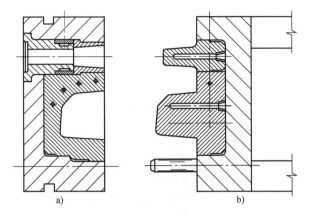

Figure 15-3   Example of insert fitting

a）Very deep cavity fitting at two levels   b）Shallow cavity

The Cores

Holes and recesses in the casting are produced by cores, usually mounted in the moving die half. Like the die inserts, they require properties which enable them to continue operating smoothly and accurately for a maximum number of shots. However, in contrast to the die insert, the replacement of cores that have become worn is not so costly, and dies that are required to operate on long runs are provided with ample spare cores especially those of small diameters. Cores are made of alloy steel, heat treated to provide maximum endurance. Where core applications call for materials with improved ductility and toughness, the 18% nickel manganese steels are specified. In brass die casting, there is a growing interest in the use of cores in molybdenum and tungsten-based refractory metals which are capable of conducting away large amounts of heat.

There are several methods for assembling cores within a die, whether the cores are fixed or moving. For our model die in Figure 15-2, the short fixed core is secured by means of a grubscrew at the rear head of the core and a second grub-screw, used for locking purposes. This arrangement enables quick core removal and replacement if damage occurs. Another method of securing short protruding cores where replacements are not often envisaged is to recess the core head position into the holding plate which is held in position by the associated die block. Flats on core heads or securing pins and keyways are often employed to prevent the cores from turning. Where several cores are required close together there may not be sufficient space for individual holding arrangements and it is common practice to secure the group of cores by means of a retaining plate.

Fixed cores are used when the axis of the cored shape is at right angles to the parting line of the die. Moving cores are required when the axis of the cored recesses are at any other angle within the die parting line. Referring to the model illustration, a moving core is actuated by means of an angled dowel pin, normally made from toughened and nitrided steel as detailed in Figure 15-4. The core is supported within a core slide, often made from chromium-vanadium steel toughened and nitrided to prevent drag or eventual seizure. Guideways for toughened slides are machined within the bolster block.

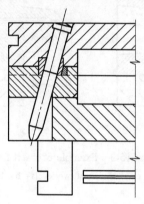

Figure 15-4   Core movement employing angled dowel pin

Lesson 16　Forging Dies

Text

　　Forging is, as far as we know, the oldest metalworking process. In the early *dawn* of civilization mankind *discovered* that a heated piece of metal was more easily hammered into different shapes[1]. Forging can be defined as the working of a piece of metal into a desired shape by hammering or pressing, usually after it has been heated to improve its plasticity. In most cases, the metal to be forged is heated to its correct forging temperature, but sometimes cold-forging is done. Cold-forging is included under cold-working and is *accomplished* in the range from room temperature up to the critical temperature of the metal.

　　General Forging Operations. In forging, the metal may be worked by

　　1. Drawing out. This increases length and decreases cross-sectional area.

　　2. *Upsetting*. This increases cross-sectional area and decreases length.

　　3. *Squeezing* in closed impression dies.

　　Superior Mechanical Properties. Forgings are generally superior in mechanical properties to castings with the same chemical analysis. Therefore, parts which must withstand severe stresses are *preferably* made by forging. Forgings have better mechanical properties than castings for at least three reasons. First, the fiber flow lines when properly controlled and directed tend to provide higher strengths. Second, the forging process by hammering or pressing produces a dense structure usually free from voids, blowholes, or *porosity*. Third, the forging process helps to refine the grain size of the metal. The working of the metal breaks up coarse grains by producing a slip along *crystallographic* planes.

　　Types of Forging Processes. The forging processes can be grouped under four principal methods as follows:

　　1. Smith forging.

　　2. Drop forging.

　　3. Press forging.

　　4. *Upset* forging.

　　Forging Design. The tools necessary to produce a given forging cannot be made until the shape of the final forging has been determined[2]. Therefore, it is essential that the tool designer has an understanding of the *underlying* principles of forging design. These will be considered in the following order: forging draft, parting planes, fillets and corner radii, *shrinkage* and die wear, mismatch of dies, tolerances, and finish allowances.

　　In smith forging a pair of dies with flat surfaces and general-purpose hand tools are used. The shaping of the part depends upon the skill of the Smith who moves the metal and directs the operation. Smith forgings are made to approximate dimensions so that machining is necessary for surfaces requiring close dimensions. This method is used for small quantities or for preliminary forging when the desired shape is to be completed by another forging method. The sizes of the forgings produced range from 1 pound to over 200 tons. Usually steam hammers, large hydraulic presses, air hammers, or helve hammers are used.

Drop-forging Dies. Drop-forging dies must withstand severe strains, resist wear, keep cracking and checking at a minimum, and have a long life under high-production conditions. In order to obtain these properties, chromium-nickel-molybdenum, chromium-nickel, or chromium-molybdenum alloy steels are used as die materials. When selecting the parting line on forgings, consideration must be given to the flow of metal and the directions of the resulting fiber flow lines. A flat parting surface in a single *horizontal* plane should be selected if possible, because irregular parting surfaces may create a side *thrust* and they add to the cost of the dies. The standard draft angle is 7 degrees since smaller draft angles cause more rapid die wear and increase the *likelihood* of the forgings sticking in the die. Interior surfaces require more draft, usually 10 degrees, since the forgings will shrink around those portions of the die as they cool if several blows are to be struck.

No-draft Forging Dies. No-draft forgings are used most often on *nonferrous* alloys. They are usually mounted in a die set and run in a forging press. Depending on the shape of the part, two or more sides of the die cavity are movable to allow the most conventional forgings and the production rates are much slower. In many cases, the parts are run once with the die not completely closed, excess flash material is removed and then the part is reheated and run a second time with the die closed completely.

Although this is an expensive process, the advantage of secondary machining operations being eliminated due to the absence of the draft angles makes the overall cost of the part considerably less.

Questions

1. What determines whether hot or cold forging is to be done?
2. What is forging draft?
3. What are the four types of forging processes?
4. What are the superior mechanical properties of forging?

New Words and Expressions

1. dawn/dɔːn/n. 黎明，曙光，开端；vi. 破晓，出现
2. discover/disˈkʌvə/vt. 看出，发现，显示，揭露；vi. 有所发现
3. accomplish/əˈkɔmpliʃ/vt. 完成，达到
4. upset/ʌpˈset/vt. 打翻，镦粗；n. 翻倒，混乱
5. squeeze/skwiːz/v. & n. 压榨，挤压
6. preferable/ˈprefərəbl/a. 更可取的，更好的
7. porosity/pɔːˈrɔsiti/n. 多孔，孔隙度，多孔结构
8. crystallographic/kristələuˈgræfik/a. 结晶的，结晶学的
9. underlying/ʌndəˈlaiiŋ/a. 在下的，基础的，根本的，潜在的

10. shrinkage/'ʃriŋkidʒ/*n*. 收缩，收缩量

11. horizontal/hɔri'zɔntl/*a*. 水平的，地平的；*n*. 地平线，水平线，地平面

12. thrust/θrʌst/*v*. 猛推，冲，插，延伸；*n*. 拉力，推力，反压力

13. likelihood/'laiklihud/*n*. 可能，相似性

14. nonferrous/'nɔn'ferəs/*a*. 非铁的

15. cold forging　冷锻

16. cross-sectional area　横截面积

17. coarse grain　粗晶粒

18. fiber flow line　纤维流线

19. nonferrous alloy　非铁合金

20. in many cases　在许多情况下

Notes

[1] In the early dawn of civilization mankind discovered that a heated piece of metal was more easily hammered into different shapes.

在人类文明的初期，人类就已经发现，加热后的金属很容易被锻打成不同的形状。

句中由 that 引出的从句作谓语动词 discovered 的宾语从句，that 只起连接作用，本身无词意。

[2] The tools necessary to produce a given forging cannot be made until the shape of the final forging has been determined.

只有当锻件的最后形状确定后，才能生产给定锻件所需的锻模。

句中 until 为连词，当主句为否定句时，until 可以先理解为"直到……以前"，然后再根据其意义译成符合汉语习惯的句子。

Glossary of Terms

1. final forging temperature　终锻温度

2. finishing temperature　终锻温度

3. initial forging temperature　始锻温度

4. finish forge　精锻，终锻

5. flat die hammer　自由锻锤

6. forge furnace　锻炉

7. forge hot（hot forging）　热锻

8. forge ratio（forge reduction）　锻造比

9. forge shop　锻造车间

10. forge tongs（weld）　锻钳（接）

11. forging burst（bursting）　锻裂

12. forging crank press　锻造用曲柄压力机

13. forging defect　锻造缺陷

14. forging die（die steel, drawing）　锻模（锻模钢，锻件图）

15. forging effect　锻造效应

16. forging line（load, practice）　锻造生产线（负荷，工艺）

17. forging plane（plant）　锻造面（厂）

18. forging pressure（process, range）　锻造力（工艺，范围）

19. free open forging　自由锻

20. impression die forging　模锻

21. open（closed）forging die　开（闭）式锻模

22. loose tooling　胎模

23. hammer forging die　锤锻模

24. forging heat-treatment　锻件热处理

25. forging temperature interval　锻造温度范围

26. forging flow line　锻造流线

27. forging tolerance　锻件公差

28. hot forging drawing　热锻件图

29. smith（hammer）forging　锤锻

30. drop forging　落锤锻造
31. press forging　压力（机）锻造
32. upsetting　镦粗，镦锻

33. heated-die forging　热模锻造
34. flashless forging　无飞边锻造
35. extrusion forging　挤压锻造

Reading Materials

Designing Forgings

Machine designers who may sometimes design forgings should fully understand the fiber-flow-line structure of steel and the benefits obtainable when the fiber flow lines are properly directed. Some of the design rules for forgings can be briefly summarized as follows:

1. When designing forgings the directions of the fiber flow lines should be kept in mind so that full advantages can be taken of them.

2. The parting lines on forgings should be in one plane if possible.

3. The parting lines on forgings should be located if possible in a plane through the center of the forging sections instead of near the upper edge, as is often done in sand casting. This is done to facilitate forging and to prevent some metal from tearing out of the forging during trimming.

4. Provision for adequate draft must be made to permit removal of the forgings from the die cavity.

5. Ample fillets and curvatures should be used because sharp corners increase the difficulty of filling die impressions and tend to promote cracks and fatigue failures.

6. Pockets and recesses should be kept at a minimum since they result in increased die wear.

7. Ribs should be made low and wide.

8. Sections should not be too thin so as to restrict the flow of metal.

9. Never specify dimensional tolerances which are closer than necessary. The ease with which dimensions can be held varies on a forging. Thickness dimensions perpendicular to the parting surface will vary some, depending upon how completely the finishing impressions close. This in turn depends upon the thickness of the flash which immediately surrounds the forging. Length and width dimensions formed completely in one die block, or die half, are easier to hold and depend largely upon shrinkage and die wear. On the other hand, length and width dimensions formed from a point in one die block to another point in the other die block are not so easy to hold, because in addition to shrinkage and die wear there is a variable in the alignment of the dies for each operation.

Forging Equipment

The most widely used type of forging equipment is the hammer, which can deform the stock with one or more sudden blows of substantial force.

Closed-die forging hammers are classified by the force that drives the weighted ram downward: either gravity or a controlled combination of gravity and compressed air. (Steam, hot air, was originally used, but such hammers are currently losing favor.)

A type of gravity hammer still in wide use, particularly for producing forgings weighing only a few pounds, is the board drop hammer. The ram is keyed to one or more wooden boards, which are then propelled upward by powered friction rolls (Figure 16-1). A trip on the ram engages a roll-release lever that moves the friction rolls away from the boards, allowing them to drop. The height of fall, adjustable mechanically, determines the force of the blow. Board hammers are rated in pounds of falling weight, the range being from 400 to 10000 lb. Air-lift gravity hammers in the same (light) range have approximately the same force.

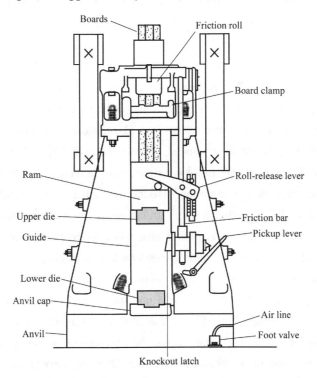

Figure 16-1　A board-type forging press (The height to which the boards are lifted determines the striking force of the gravity hammer.)

The largest forging presses in the world are hydraulic-powered forging presses. More recently NC has been made available for hydraulic forging presses so that any sequence can be programmed, such as rapid approach of the ram, and then decelerate as contact is made, all the while maintaining just the right velocity to make the metal flow at the desired rate. After a controlled dwell period, the ram is driven up quickly. High-energy-rate forging (HERF) presses are also used in forging.

Unit Six

Lesson 17   Plastics

Text

Plastic means *pliable* or impressionable. It is defined as the capability of being deformed continuously in any direction without breaking apart[1]. Following this *definition*, plastics could include glass, metals, and *wax*. Thus, it is evident that the name plastics is misleading and not exactly correct. Today, however, the name plastics is identified with the products which are derived from *synthetic resins*. The synthetic resins are made by various chemical processes. Plastics manufacturing is of comparatively recent date. The discovery of *ebonite*, or hard *rubber*, by Charles Goodyear in 1839, and the development of *celluloid* (*cellulose nitrate*) by Hyatt in 1869 marked the beginning of plastic products. It was not until 1909, however, that one of the most important materials, *phenol formaldehyde* resin, was developed by Dr. L. H. Baekeland and his associates. Today plastics are commonly found in homes, automobiles, numerous other products, and machines. Some machines which utilize electricity require plastic parts with electrical insulating properties. One of the largest plastic products ever built is a 2600-pound 28-foot-long boat *hull* which is made of fiber-glass matting *impregnated* with *polyester* resin and low-pressure *laminated*. The unusual properties of some plastic materials are used in many interesting applications. Advantage should be taken of these unusual properties wherever possible.

Thermosetting or *Thermoplastic*. Plastic materials, commonly called plastics, are known either as thermosetting or as thermoplastic. Upon the application of initial heat the thermosetting plastics soften and melt, and under pressure they will fill every *crevice* of a mold cavity. Upon continued application of heat they *polymerize*, or undergo a chemical change, which hardens them into a permanently hard, *infusible*, and insoluble state. After this they cannot again be softened or melted by further heating. The thermoplastic materials are those which soften with heating and solidify and harden with cooling. They can be remelted and cooled time after time without undergoing any *appreciable* chemical change.

Fillers. Plastics often contain other added materials called fillers. Fillers are employed to increase bulk and to help *impart* desired properties. Plastics containing fillers will cure faster and hold closer to established finished dimensions, since the plastic shrinkage will be reduced. Wood flour is the general-purpose and most commonly used filler. Cotton frock, produced from cotton *linters*, increases mechanical strength. For higher strength and resistance to impact, cotton cloth *chopped* into sections about 1/2-inch square can be processed with the plastic. *Asbestos* fiber may be used as a filler for increased heat and fire resistance, and *mica* is used for molding plastic parts with superior *dielectric* characteristics[2]. Glass fibers, silicon, cellulose, clay, or nutshell flour may also be used. Nutshell flour is used instead of wood flour where a better finish is desired. Plastic parts using short fiber fillers will result in lower costs, while those with long fiber fillers having greater impact strengths are more expensive. Other materials, not defined as fillers, such as dyes, *pigments*, lubricants, accelerators, and plasticizers may also be added. Plasticizers are added to soften and improve the moldability of plastics. Filler and modifying agents are added and mixed with the raw plastic before it is molded or formed.

Properties. Numerous plastics have already been discovered and developed by the chemist and chemical engineer, and research on the synthetic resins is one of the most *prominent* fields in organic chemistry today. The mechanical properties are considerably lower than those for metals with the tensile strengths usually ranging from 5, 000 to 15, 000 psi.

Some of the plastics listed, such as phenol formaldehyde resin, consist of several different varieties and mixtures included under the one family. Numerous other plastics with special properties are available. However, the ones listed should serve as a brief introduction.

Questions

1. What does plastic mean?
2. List some applications of the plastics.
3. Describe the differences between thermosetting and thermoplastic.
4. Describe the function of fillers in plastics.

New Words and Expressions

1. plastic/ˈplæstik/ *n.* 塑料，塑料制品；*a.* 塑性的，塑料的

2. pliable/ˈplaiəbl/ *a.* 柔韧的，可锻的，可塑的

3. definition/defiˈniʃən/ *n.* 定义，解说，限定，清晰度

4. wax/wæks/ *n.* 蜡；*a.* 蜡制的；*vi.* 给……上蜡

5. synthetic/sinˈθetik/ *a.* 合成的，人造的，综合的；*n.* 合成剂

6. resin/ˈrezin/ *n.* 树脂，松香，树脂制品；*vt.* 涂树脂，用树脂处理

7. ebonite/ˈebənait/ *n.* 胶木，硬橡胶

8. rubber/ˈrʌbə/ *n.* 橡皮，橡胶，合成橡胶；*vt.* 给……涂上橡胶

9. celluloid/ˈseljulɔid/ *n.* （物）赛璐珞，明胶；*a.* 细胞状的

10. cellulose/ˈseljuləus/ *n.* 纤维素

11. nitrate/ˈnaitreit/ *n.* 硝酸盐；*vt.* 使硝化

12. phenol/ˈfiːnɔl/ *n.* 酚，碳酸

13. formaldehyde/fɔːˈmældihaid/ *n.* 甲醛

14. hull/hʌl/ *n.* 船体

15. impregnate/ˈimpregneit/ *vt.* 使充满，使饱和

16. polyester/pɔliˈestə/ *n.* 聚酯

17. laminate/ˈlæmineit/ *vt.* 切成薄片，层压；*n.* 层状，层压，材料

18. thermosetting/θəːməuˈsetiŋ/*n.* & *a.* 热固性（的）

19. thermoplastic/θəːməuˈplæstik/*a.* 热塑性的；*n.* 热塑性塑料

20. crevice/ˈkrevis/*n.* 裂缝

21. polymerize/ˈpɔliməraiz/*v.* （使）聚合；*n.* 聚合，重合

22. infusible/inˈfjuːzəbl/*a.* 难熔化的，不熔化的

23. appreciable/əˈpriːʃiəbl/*a.* 可估计的，明显的，可观的

24. filler/ˈfilə/*n.* 填料，填充物

25. impart/imˈpɑːt/*vt.* 分给，给予，授予

26. linter/lintə/*n.* 绒布，棉绒，短棉纤维

27. chop/tʃɔp/*vt.* *vi.* 斩，斩碎；*n.* 切断，品种

28. asbestos/æzˈbestəs/*n.* 石棉

29. mica/ˈmaikə/*n.* 云母

30. dielectric/daiiˈlektrik/*n.* 电介质，绝缘材料

31. pigment/pigmənt/*n.* 色素，颜料

32. prominent/ˈprɔminənt/*a.* 杰出的，突出的

33. cellulose nitrate 硝酸纤维

34. synthetic resin 合成树脂

35. polyester resin 聚酯树脂

36. phenol-formaldehyde resin 酚醛树脂

Notes

[1] Plastic means pliable or impressionable. It is defined as the capability of being deformed continuously in any direction without breaking apart.

所谓塑性，是指材料是柔韧的或可塑的。根据定义，塑性材料具有在任何方向上连续变形而不发生断裂的能力。

句中 being deformed（一般式的被动形式）是动名词短语作介词 of 的宾语，动词 deform 为"变形"之意；in any direction 译为"沿任一方向"，这里介词 in 意为"朝，向"等，例如：in the same direction 沿相同方向，in all directions 四面八方。

[2] Asbestos fiber may be used as a filler for increased heat and fire resistance, and mica is used for molding plastic parts with superior dielectric characteristics.

用石棉纤维作为填料可以增强塑料的耐热性和耐燃性，而用云母可使塑料件具有良好的电绝缘性能。

句中介词短语 with superior dielectric characteristics 作定语，修饰 parts 一词，可译为"具有优越的电绝缘性能"。

Glossary of Terms

1. foamed (cellular) plastic 泡沫塑料
2. thermosetting plastic 热固性塑料
3. plastics industry 塑料（工业）行业
4. blow mold for plastics 塑料吹塑模
5. standard mold components for plastics 塑料模具标准化零部件
6. thermoforming machine for plastics 塑料热成型机
7. mold for foamed plastics 泡沫塑料模型
8. plastic molding press 塑料压力成型机
9. extruder double-screw for plastics 塑料加工用双螺杆挤出机
10. extruder single-screw for plastics 塑料加工用单螺杆挤出机

11. laminating machine　层压机
12. polymethyl methacrylate（PMMA）　聚甲基丙烯酸甲酯（有机玻璃）
13. acrylonitrile-butadiene-styrene plastic（ABS）苯乙烯-丁二烯-丙烯腈塑料
14. polyvinyl chloride（PVC）　聚氯乙烯
15. polyamide（PA）　聚酰胺
16. polyethylene terephthalate（PET）　聚对苯二甲酸乙二酯
17. polypropylene（PP）　聚丙烯
18. polystyrene（PS）　聚苯乙烯
19. polyethylene（PE）　聚乙烯
20. polycarbonate（PC）　聚碳酸酯
21. polyoxymethylene（POM）　聚甲醛
22. epoxy resin（EP）　环氧树脂
23. nylon　尼龙
24. engineering plastic　工程塑料
25. biodegradable plastic　生物降解塑料

Reading Materials

Preforms

In mass-production compression molding or transfer molding, it would require too much of the operator's time to weigh out individually each charge of powdered thermosetting material. For this reason the thermosetting material usually is cold-pressed in a preliminary operation into preforms of various shapes and sizes. Sufficient pressure is used so the preforms can be handled, but no polymerization takes place. The preforms are accurately measured and the operator quickly loads one or more preforms into the mold cavity. Without preforming, it would be practically impossible to operate large semiautomatic molds at mass-production rates. The preforming operation consists of cold compressing the plastic powder into tablets of correct size and weight in a press specially designed for that purpose. Since large quantities are needed, preforms are often made in automatic rotary presses. These presses have 12 or more similar molds equally spaced around the rotary table. The powder is automatically fed from a hopper to the molds as they pass the loading station. Each mold has both an upper and a lower punch which move vertically. At the loading station the upper punch is up clear of the mold while the lower punch, which forms the bottom of the mold cavity, is down at its lowest position. As the table continues to rotate, the excess plastic powder is scraped off level with the top of the mold. Next the upper punch descends into the mold and both punches are forced together, compressing the powder. To eject the preform, the upper punch rises clear of the mold again while the lower punch rises to the top level of the mold so the preform is entirely ejected. An arm guides the preform off the rotating table where it slides down a chute into an appropriate container. The mold moves onto the loading station and the cycle starts again.

Thermoplastic Elastomers

Thermoplastic elastomers are similar to the thermosetting type in service properties. However, they do offer greater ease and speed of processing. Molding or extruding can be done on standard plastic-processing equipment with considerably shorter cycle times than that required for compression molding of conventional rubbers.

The principal thermoplastic elastomers are polyester copolymers, styrene-butadiene block copolymers, and polyurethane, from a number of producers.

Polyurethane. Polyurethane is the first major elastomer to be developed that could be processed as a thermoplastic. It does not have quite the heat resistance or the compression set resistance of the crosslinked types, but most other properties are similar.

Material Specifications and Properties of Elastomers. The system of specifying elastomers as given in ASTM D2000 or SAE 5200 is based on properties required, rather than material composition. An example of a specification requirement may be as follows:

Applicable type of material:	chloroprene
Original properties:	
Hardness	50 ±5 (durometer)
Tensile	1500 psi (10.34MPa) min
Elongation	350% min
Air oven, 70 hr at 100℃:	
Hardness change	+15max
Tensile change	−15max
Elongation change	−40% max
ASTM #3 oil, 70 hr at 100℃:	
Tensile change	−70% max
Elongation change	−55% max
Volume change	+120% max
Compression set, 22 hr at 100℃:	
Set	80% max

Lesson 18   Compression Molding for Plastics

Text

In *compression* molding the plastic material as *powder* or preforms is placed into a heated steel mold cavity. Since the parting surface is in a horizontal plane, the upper half of the mold descends vertically. It closes the mold cavity and applies heat and pressure for a *predetermined* period. A pressure of from 2 to 3 tons per square inch and a temperature at approximately 350 °F converts the plastic to a semiliquid which flows to all parts of the mold cavity. Usually from 1 to 15 minutes is required for curing, although a recently developed *polyester* plastic will *cure* in less than 25 seconds. The mold is then opened and the molded part removed. If metal inserts are desired in the parts, they should be placed in the mold cavity on pins or in holes before the plastic is loaded[1]. Also, the preforms should be preheated before loading into the mold cavity to eliminate gases, improve flow, and decrease curing time. Dielectric heating is a *convenient* method of heating the preforms.

Since the plastic material is placed directly into the mold cavity, the mold itself can be simpler than those used for other molding processes. Gates and sprues are unnecessary. This also results in a saving in material, because *trimmed*-off gates and sprues would be a complete loss of the thermosetting plastic. The press used for compression molding is usually a vertical hydraulic press. Large presses may require the full attention of one operator. However, several smaller presses can be operated by one operator. The presses are conveniently located so the operator can easily move from one to the next. By the time he gets around to a particular press again, that mold will be ready to open.

The thermosetting plastics which harden under heat and pressure are suitable for compression molding and transfer molding. It is not practical to mold thermoplastic materials by these methods, since the molds would have to be alternately heated and cooled. In order to harden and *eject* thermoplastic parts from the mold, cooling would be necessary[2].

Types of molds for compression molding. The molds used for compression molding are classified into four basic types, namely, positive molds, landed positive molds, *flash*-type molds, and semipositive molds. In a positive mold the *plunger* on the upper mold enters the lower mold cavity. Since there are no lands or stops on the lower mold, the plunger completely *traps* the plastic material and descends with full pressure on the charge. A dense part with good electrical and physical properties is produced. The amount of plastic placed in the mold cavity must be accurately measured, since it determines the thickness of the part. A landed positive mold is similar to a positive mold except that lands are added to stop the travel of the plunger at a predetermined point. In this case, the lands absorb some of the pressure that should be exerted on the parts. The thickness of the parts will be accurately controlled, but the density may vary considerably. In a flash-type mold, flash *ridges* are added to the top and bottom molds. As the upper mold *exerts* pressure on the plastic, excess material is forced out between the flash ridges where it forms flash. This flash is further compressed, becomes hardened, and finally stops the downward travel of the upper mold. A slight excess of plastic material is always charged to ensure sufficient pressure to produce a dense molded part. This type of mold is widely used because it is comparatively easy to construct and it controls thickness and density within close limits[3]. The semipositive mold is a combination of the flash type and landed positive molds. In addition to the flash ridges, a land is employed to *restrict* the travel of the upper mold.

Questions

1. What is the compression molding of the plastics?
2. Describe the flash-type mold.
3. What are the advantages of the compression molding?
4. How many types of molds are there for compression molding?

New Words and Expressions

1. compression/kəm'preʃən/*n.* 压紧，压缩，压力
2. powder/'paudə/*n.* 粉末，粉剂，药粉；*v.* 磨成粉
3. predetermine/'priːdi'təːmin/*vt.* 预定
4. polyester/pɔli'estə/*n.* 聚酯
5. cure/kjuə/*n.* 治愈，医治，熟化，固化；*vt.* 纠正，治疗
6. convenient/kən'viːnjənt/*a.* 方便的，便利的，合宜的，附近的
7. trimmed/trimd/*a.* 平衡的；修整过的
8. eject/i'dʒekt/*vt.* 排斥，喷射，弹射，抛出
9. flash/flæʃ/*n.* 闪光，一瞬间，溢料，毛刺；*a.* 突然出现的
10. plunger/'plʌndʒə/*n.* 活塞，柱塞，潜水者，短路器
11. trap/træp/*n.* 陷阱，捕机，活门，闸门；*vt.* 收集，止住
12. ridge/ridʒ/*n.* 山脉，波峰，背；*vt.* 使起皱
13. exert/ig'zəːt/*vt.* 行使，运行，尽（力），施加（力）
14. restrict/ri'strikt/*vt.* 限制，限定，约束
15. flash-type mold 溢式压缩模
16. trimmed-off 清理，去掉
17. per square inch 每平方英寸
18. horizontal plane 水平面
19. mold cavity 模具型腔
20. classify...into... 把……分类为

Notes

［1］If metal inserts are desired in the parts, they should be placed in the mold cavity on pins or in holes before the plastic is loaded.

如果零件中需要金属镶嵌件时，应在塑料加载前将这些嵌件放入型腔定位柱或定位孔中。

if 引出条件状语从句，译为"如果……"；主句中 they 代表 metal inserts。

［2］In order to harden and eject thermoplastic parts from the mold, cooling would be necessary.

为了使热塑性塑件硬化及将其从型腔中顶出, 就需要将塑件冷却。

句中 in order to 引导目的状语从句, 译为"为了使……"。

［3］This type of mold is widely used because it is comparatively easy to construct and it controls thickness and density within close limits.

这种类型的模具之所以在生产中被广泛采用, 是因为它相对比较容易制造, 并能够在极限范围内控制塑件的厚度和密度。

句中 easy to construct 属于"形容词 + 不定式"这一结构, 可译为"容易制造", 又如 difficult to solve 可译为"难以解决"。

Glossary of Terms

1. compression mold 压缩模
2. parting surface 分型面
3. transfer mold 压注模 (也称传递模)
4. flash-type mold 溢式压缩模
5. semi-positive mold 半溢式压缩模
6. positive mold 不溢式压缩模
7. portable transfer mold 移动式压注模
8. thermosetting resin binder 热固性树脂黏结剂
9. dwell pressure 保压压力
10. dwell time 保压时间
11. filling velocity 充模速度
12. mold for plastics 塑料成型模具 (简称塑料模)
13. mold for thermoplastics 热塑性塑料模
14. draft 脱模斜度
15. fixed compression mold 固定式压缩模
16. mold opening force 开模力
17. mold pressure 成型压力
18. sprue puller 拉料杆
19. core-pulling force 抽芯力
20. injection speed 注射速度
21. parting line 分型面
22. plunger unit 柱塞装置

Reading Materials

Plastic Processing

There are two main steps in the manufacturing of plastic products. The first is a chemical process to create the resin. The second is to mix and shape all the material into the finished article or product.

Plastic objects are formed by compression, transfer, and injection molding. Other processes are casting, extrusion and laminating, filament winding, sheet forming, joining, foaming, and machining. Some of these and still others are used for rubber. A reason for a variety of processes is that different materials must be worked in different ways. Also, each method is advantageous for certain kinds of products. The principles of operation and merits of the processes will be discussed.

Compression Molding. In compression molding a proper amount of material in a cavity of a mold is squeezed by a punch, also called a force. The plastic is heated in most cases between 120 and 260°C (250 and 500°F), softens, and flows to fill the space between force and mold. The mold is kept closed for enough time to permit the formed piece to harden. This is done in a press capable of exerting 15 to 55 MPa (2000 to 8000 psi) over the area of the work projected on a plane normal to the ram movement, depending on the design of the part and the material.

Compression molding is mostly for thermosetting plastics which have to be cured by heat in the mold. Other methods are faster for large-quantity production of thermoplastics. Loose molding compound may be fed into a mold, but a cold-pressed tablet or rough shape, called a preform, may be prepared for more rapid production. For efficient heat transfer, parts should be simple with walls uniform and preferably not over 3mm (1/8in) thick. Even so, it may take several minutes to heat and cure a charge. This time may be reduced as much as 50% by preheating the charge. To speed the process as much as possible, molding presses are usually semi- or fully automatic.

The three basic types of compression molds for plastics are shown in Figure 18-1. The force fits snugly in the positive-type mold. The full pressure of the force is exerted to make the material fill out the mold. The amount of charge must be controlled closely to produce a part of accurate size.

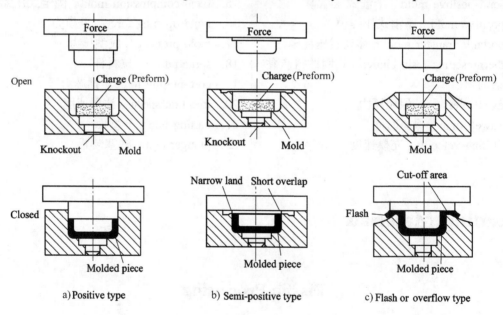

Figure 18-1   Three types of compression molds for plastics

The force is a close fit in a semipositive-type mold only within the last millimeter of travel. Full pressure is exerted at the final closing of the mold, but excess material can escape, and the charge does not have to be controlled so closely. This type is considered best for large-quantity production of pieces of quality.

The force does not fit closely but closes a flash- or overflow-type mold by bearing on a narrow flash ridge or cutoff area. The amount of material does not need to be controlled closely, and the excess is squeezed out around the cavity in a thin flash. Some material is wasted, and all pieces must be trimmed. Full pressure is not impressed on the workpiece. A mold of this kind is usually cheapest to make.

Blow Molding

The blow-molding process produces the ubiquitous designer water bottle and similar consumer products. Extrusion blow molding is shown in Figure 18-2, and similar processes include injection blow molding and stretch blow molding.

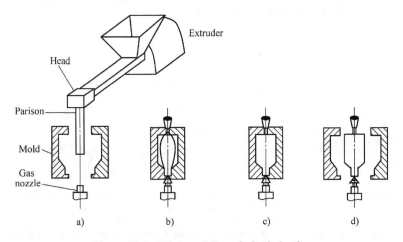

Figure 18-2　Blow molding of plastic bottles

The extruder first extrudes a heated dangling hollow tube, called a parison, into the open mold. The mold closes and pinches off the upper tube end, this usually corresponds to the bottom of the bottle. The reader might glance at any plastic drinking bottle and see the rough nipple that gets created. In some designs a crease may be used. Either way, a new problem is created for the part designer! The rough projection on the bottom will make the bottle unstable as it sits on a user's desk. Another glance at the bottom of any bottle will thus show a carefully designed convex hull or some feet like projections around the bottle's perimeter.

After the mold closes, the parison is inflated and blown out to adopt the shape of the mold. Note that the extruded parison is already heated and the mold opening is warmed so that the leathery polymer easily takes up its molded shape. Subsequently, the now formed bottle is cooled and removed from the mold. A new parison is then extruded into the open mold.

Lesson 19   Transfer Molding and Injection Molding

Text

Transfer Molding. In compression molding where preforms are compressed and reduced to a fluid in the mold cavity, it is possible for *slender* cores to be broken and for inserts to be *dislodged*. Also, for complex parts with thin sections, the proper flow of plastic is difficult to obtain. To overcome these difficulties transfer molding has been developed. In transfer molding the plastic material is forced into the mold cavity as a fluid. It flows freely around inserts and slender cores and fills the mold cavity (Figure 19-1).

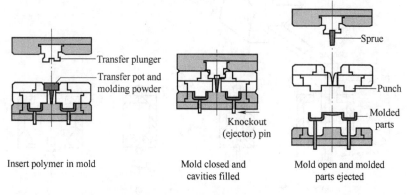

Insert polymer in mold   Mold closed and   Mold open and molded
                             cavities filled   parts ejected

Figure 19-1   Sequence of operation in transfer molding for thermosetting plastics

Basically, there are two types of transfer molds, the *conventional sprue* type and the positive plunger type. In the sprue type the plastic preforms are placed in a separate loading chamber above the mold cavity. One or more sprues lead down to the parting surface of the mold where they connect with gates to the mold cavity or cavities. Special presses with a floating *intermediate platen* are especially useful for *accommodating* the two parting surface molds. The plunger acts directly on the plastic material, forcing it through the sprues and gates into the mold cavities. Heat and pressure must be maintained for a definite time for curing. When the part is cured the press is opened, breaking the sprues from the gates. The *cull* and sprues are raised upward, being held by a tapered, *dovetailed* projection machined on the end of the plunger. They can easily be removed from the dovetail by pushing horizontally. In a positive plunger-type transfer mold the sprue is eliminated so that the loading chamber extends through to the mold parting surface and connects directly with the gates. The positive plunger type is preferred, because the mold is less complicated, and less material is wasted. Parts made by transfer molding have greater strengths, more uniform densities, closer dimensional tolerances, and the parting line requires less cleaning as compared with parts made by compression molding[1].

Injection Molding. Injection molding is principally used for the production of thermoplastic parts, although some progress has been made in developing a method for injection molding some thermosetting materials. The problem of injecting a melted plastic into a mold cavity from a *reservoir* of melted material has been extremely difficult to solve for thermosetting plastics which cure and harden under such conditions within a few minutes. The principle of injection molding is quite similar to that of die-casting. Plastic powder is loaded into the feed *hopper* and a certain amount feeds into the heating chamber when the plunger draws back. This plastic powder under heat and pressure in the heating chamber becomes a fluid. Heating temperatures range from 265 to 500 ℉. After the mold is closed, the plunger moves forward, forcing some of the fluid plastic into the mold cavity under pressures ranging from 12000 to 30000 psi. Since the mold is cooled by circulating cold water, the plastic hardens and the part may be ejected when the *plunger* draws back and the mold opens. Injection-molding machines can be arranged for manual operation, automatic single-cycle operation, and full automatic operation. Typical machines produce molded parts weighing up to 22 ounces at the rate of four *shots* per minute, and it is possible on some machines to obtain a rate of six shots per minute[2]. The molds used are similar to the dies of a die-casting machine with the exception that the surfaces are *chromium*-plated. The advantages of injection molding are:

1. A high molding speed adapted for mass production is possible.

2. There is a wide choice of thermoplastic materials providing a variety of useful properties.

3. It is possible to mold *threads*, undercuts, side holes, and large thin sections.

Questions

1. What difficulties can be overcome by transfer molding?

2. Describe two types of transfer molds.

3. What is the difference between the positive plunger type and the conventional sprue type?

4. What are the advantages of injection molding?

New Words and Expressions

1. slender/ˈslendə/*a.* 细长的，不足的，微小的

2. dislodge/disˈlɔdʒ/*vt.* 移去，移位，取出，移动

3. conventional/kənˈvenʃənl/*a.* 常规的，传统的

4. sprue/spruː/*n.* 浇口，流道

5. intermediate/intəˈmiːdjət/*a.* 中间的，居中的；*n.* 中间人，中间体

6. platen/ˈplætən/*n.* 台板，隔板，压板

7. accommodate/əˈkɔmədeit/*vt.* 容纳，接纳，供给，提供

8. cull/kʌl/*vt.* 采，摘，选拔；*n.* 拣剩的东西

9. dovetail/ˈdʌvteil/*n.* 楔形榫，鸠尾榫；*v.* 吻合

10. injection/inˈdʒekʃən/*n.* 注入，注射，射入，加压充满

11. reservoir/ˈrezəvwɑː/*n.* 容器，储存器；*vt.* 储藏

12. hopper/ˈhɔpə/*n.* 漏斗，给料斗，计量器

13. plunger/ˈplʌndʒə/*n.* 柱塞，活塞
14. shot/ʃɔt/*n.* 注射；发射
15. chromium/ˈkrəumjəm/*n.* 铬
16. thread/θred/*n.* 线（状物），丝，螺纹；*vt.* 插入，车螺纹，攻螺纹

17. be similar to　类似（相似）于……的
18. as compared with　与……相比
19. injection molding　注射模
20. dimensional tolerance　尺寸公差

Notes

[1] Parts made by transfer molding have greater strengths, more uniform densities, closer dimensional tolerances, and the parting line requires less cleaning as compared with parts made by compression molding.

与压缩模塑相比，传递模塑生产出的塑件不仅具有较高的强度、比较均匀的密度及较为精确的尺寸，而且分型面上需要清理的毛刺也少。

句首的 made by transfer molding 为分词短语作后置定语修饰 parts 一词，意为"用传递模生产出的塑件"；句尾的 parts made by compression molding 也属于此类情况。

[2] Typical machines produce molded parts weighing up to 22 ounces at the rate of four shots per minute, and it is possible on some machines to obtain a rate of six shots per minute.

典型机器每分钟注射 4 次，塑件质量可达 22 盎司，在一些机器上可以获得每分钟注射 6 次的速率。

句中的 weighing up to 意为"重达……"，这是一个分词短语作后置定语，被修饰的词是 parts；at the rate of... 意为"以……的速率"。

Glossary of Terms

1. transfer mold　压注模（传递模）
2. injection mold　注射模
3. injection mold for thermosets　热固性塑料注射模
4. injection mold for thermoplastics　热塑性塑料注射模
5. portable transfer mold　移动式传递模
6. fixed transfer mold　固定式传递模
7. runnerless mold　无流道模
8. hot runner mold　热流道模
9. insulated runner mold　绝热流道模
10. warm runner mold　温流道模
11. ring gate　环形浇口
12. pin-point gate　点浇口
13. edge gate　侧浇口
14. submarine gate, tunnel gate　潜伏浇口
15. cold-slug well　冷料穴
16. sprue bush　浇口套
17. runner plate　流道板
18. spreader　分流锥
19. warm runner plate　温流道板
20. stationary mold, fixed half cover die　定模
21. moving mold, moving half　动模
22. shot volume　注射量，压注量
23. direct gate　直浇口
24. positive plunger type　正柱塞型
25. primary sprue　主流道
26. locating ring　定位环

Reading Materials

Typical Two-Plate Injection Mold

Figure 19-2 shows an exploded view of the elements of a typical two-plate injection mold. A brief description of each element as follows:

1. Return pin—returns the ejector plate back to its original position when the mold is closed.

2. Ejector pin—pushes part off the core or out of the cavity.

3. Sprue puller pin—pulls sprue out of the bushing when mold opens by means of an undercut (not shown).

4. Ejector retainer plate—plate in which the ejector pins are held.

5. Ejector plate—often referred to as the ejector cover plate. Provides backup for pins set into the ejector-retaining plate.

6. Support pillar—gives strength and rigidity for the ejector plates.

7. Ejector housing—provides travel space for the ejector plate and ejector pins.

8. Support plate—adds rigidity and strength to the plate stackup (not used in all molds).

9. Core retainer plate—holds the core element, the mating half of the cavity.

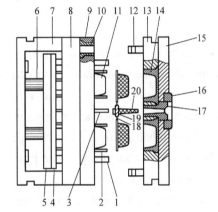

Figure 19-2   Conventional two-plate mold

10. Leader pin bush—provides close tolerance guide for leader pin.

11. Core—the male part of the cavity.

12. Leader pin—provides alignment for the two halves of the mold as it opens and closes.

13. Cavity retainer plate—holds the cavity inserts (the cavity is the area where the part is formed).

14. Cavity—place where the plastic is formed. Some molds may have the cavity cut directly in the cavity plate rather than by the use of an insert as shown.

15. Clamp plate—secures the stationary side of the mold to the molding machine. Clamps are inserted in the recess between the clamp plate and the cavity-retainer plate.

16. Locating ring—aligns the nozzle of the injection-molding machine with the mold.

17. Sprue bush—a conical channel that carries the injected plastic through the top clamp plate to the part or runner system.

18. Gate—the restricted area of the runner right before the material enters the cavity.

19. Runner—a passageway for the plastic to flow from the sprue to the part.

20. Sprue—a passageway for the plastic from the nozzle to the runner.

Unit Seven

Lesson 20　Fundamentals of Chip–Type Machining Processes

Text

Machining is the process of removing unwanted material from a workpiece in the form of *chips*. If the workpiece is a metal, the process is often called metal cutting or metal removal. U. S. industries annually spend $ 60 billion to perform metal removal operation because the vast majority of manufactured products require machining at some stage, their production, ranging from relatively rough or nonprecision work, such as cleanup of castings or forgings, to high-precision work involving *tolerances* of 0.0001in or less and high-quality finishes[1]. Thus machining undoubtedly is the most important of the basic manufacturing processes.

Over the past 80 years, the process has been the object of considerable research and experimentation that have led to improve understanding of the nature of both the process itself and the surfaces produced by it[2]. Although this research effort has led to improvements in machining *productivity*, the complexity of the process has resulted in slow progress in obtaining a complete theory of chip formation.

What makes this process so unique and difficult to analyze?

Different materials behave differently.

The process is *asymmetrical* and *unconstrained*, *bounded* only by the cutting tool.

The level of *strain* is very large. The strain rate is very high.

The process is *sensitive* to variations in tool geometry, tool material, temperature, environment (cutting fluids) and process *dynamics* (chatter and vibration).

The objective of this chapter is to put all this in *perspective* for the practicing engineer.

There are seven basic chip formation processes: shaping, turning, milling, drilling, sawing, *broaching*, and grinding (*abrasive* machining). For all metal-cutting processes, it is necessary to distinguish between speed, feed, and depth of cut. The turning process will be used to introduce these terms (see Figure 20-1). In general, speed

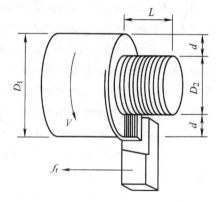

Figure 20-1　Relationship of speed, feed, and depth of cut in turning

(V) is the primary cutting motion, which relates the velocity of the rotating workpiece with respect to the stationary cutting tool[3]. It is generally given in units of surface feet per minute or inches per minute (in/min), or meters per minute (m/min), or meters per second (m/s). Speed (V) is shown in Figure 20-1 with the heavy dark arrow. Feed (f_r) is the amount of material removed per revolution or per pass of the tool over the workpiece. In turning feed is in inches/revolution and the tool feeds parallel to the rotational axis of the workpiece. Feed units are inches per revolution, inches per minute, or inches per tooth, depending on the process. Feed is shown with *dashed* arrows. The depth of cut (DOC), d, represents the third dimension. In turning, it is the distance the tool is *plunged* into the surface. It is half the difference in the diameters D_1, the initial diameter, and D_2, the final diameter[4]:

$$d = \frac{D_1 - D_2}{2} \qquad\qquad (20\text{-}1)$$

Speed, started in surface feet per minute, is the peripheral speed at the cutting edge. Feed per revolution in turning is a linear motion of the tool parallel to the rotating axis of the workpiece. The depth of cut reflects the third dimension. L represents the length of cut.

The surface speed of the rotating part is related to the outer diameter of the workpiece:

$$V = \frac{\pi D_1 N}{12} \qquad\qquad (20\text{-}2)$$

where D_1 —the initial diameter, inch;

V—the speed in surface feet per minute;

N—the revolutions per minute (rpm) of the workpiece.

Another figure (abridged by editor) shows a typical machine tool for the turning process, a lathe. Workpieces are held in workholding devices. In this example, a three-jaw *chuck* is used to hold the workpiece and rotate it against the tool. The chuck is attached to the *spindle*, which is driven through gears by the motor. The cutting tool is used to machine the workpiece and is the most *critical* component. Cutting tool material and geometry must be selected before speed and feed can be determined. Machine tools, cutting tools, and workholding devices are usually manufactured by separate companies and the tooling can cost as much or more than the machine tool.

Questions

1. What does the term *metal cutting* mean? Why does the U. S. industry spend $ 60 billion for it every year?
2. Why did the process of metal cutting progress slowly in the past 80 years?
3. What are the most important factors for all metal cutting processes?
4. Are the finishing cuts heavier than roughing cuts in terms of feed and DOC, and with a high surface speed? True or false? Tell where can you find the answer.

New Words and Expressions

1. chip/tʃip/ *n.* 碎屑；*v.* 切削（切成小碎片）
2. tolerance/ˈtɔlərəns/ *n.* 容忍；公差
3. productivity/prɔdʌkˈtiviti/ *n.* 生产率，生产能力
4. asymmetrical/æsiˈmetrikəl/ *a.* 不平衡的，不对称的
5. constrained/kənˈstreind/ *a.* 被强迫的，受约束的
6. bound/baund/ *vt.* *a.* & *n.* 受约束（的）
7. strain/strein/ *n.* 应变；*v.* 拉紧
8. sensitive/ˈsensitiv/ *a.* 敏感的
9. dynamic/daiˈnæmik/ *a.* 动力的；*n.* 动力学
10. perspective/pəˈspektiv/ *n.* 透视，一个
11. broach/brəutʃ/ *n.* 拉（削）刀；*v.* 拉削，扩（铰）孔
12. abrasive/əˈbreisiv/ *n.* 磨料；*a.* 有研磨作用的
13. dash/dæʃ/ *n.* 急奔，破折号，dash arrow 长箭头
14. plunge/plʌndʒ/ *vt.* & *vi.* -（into）投入，插入
15. chuck/tʃʌk/ *n.* 夹头，the three-jaw chuck 三爪卡盘
16. spindle/ˈspindl/ *n.* （主）轴
17. critical/ˈkritikəl/ *a.* 临界的，苛刻的，要求高的

问题不同方面的关系

Notes

[1] U. S. industries annually spend $ 60 billion to perform metal removal operation because the vast majority of manufactured products require machining at some stage, their production, ranging from relatively rough or nonprecision work, such as cleanup of castings or forgings, to high-precision work involving tolerances of 0.0001in or less and high-quality finishes.

美国工业界每年耗资 600 亿美元用于完成金属切削加工，这是因为大量的产品在制造阶段需要进行切削加工。生产加工范围从相对粗糙或没有精度要求的产品，如铸件和锻件的去黑皮加工，到公差为 0.0001 英寸（2.54μm）或更小的高精度和高表面质量的产品。

because 引导原因状语从句，说明耗巨资的原因；分词短语 ranging from...to... 进一步说明产品的范围，意为："范围从……到……"；vast 原意为"巨大的"，majority 意为"大多数"，在这里合译为"大量的"；cleanup 原意为"清洗"，这里借意为"去除"。

[2] Over the past 80 years, the process has been the object of considerable research and experimentation that have led to improve understanding of the nature of both the process itself and the surfaces produced by it.

在过去的 80 年里，人们通过对金属切削加工工艺的大量研究和试验，增进了对加工工艺性能和被加工表面特性的了解。

that 引导说明主句中研究和试验的结果，that 代表整个主句，而主句省略了动作的主体"人们"，直接以 process 为主语。

〔3〕 In general, speed（*V*）is the primary cutting motion, which relates the velocity of the rotating workpiece with respect to the stationary cutting tool.

通常，切削速度（*V*）是指切削时的主运动速度，它表达旋转的工件相对于固定不动的刀具的速度。

应注意固定搭配 with respect to..., 译为"关于……"，which 代表速度 *V*。

〔4〕 It is half the difference in the diameters D_1, the initial diameter, and D_2, the final diameter.

它是工件初始直径 D_1 和加工后直径 D_2 差值的一半。

原句相当于"It is half of the difference between the diameters D_1, the initial diameter, and D_2, the final diameter."在理解上需要注意。

Glossary of Terms

1. chip pocket　容屑槽
2. chip formation　成形切削
3. chip load（force）　切削力
4. drilling and reaming　钻孔和铰孔
5. taper turning　锥度车削
6. external threading　外螺纹车削
7. single-point cutting tool　单尖切削刀具
8. chuck handle　夹头扳手，夹头钥匙
9. combination chuck　复动夹头
10. independent chuck　分动夹头（四爪夹头）
11. union chuck　双爪夹头
12. drill jig　钻套，钻夹具
13. grinding wheel　砂轮
14. shaper and planer　牛头刨床和龙门刨床
15. slotting machine　插床
16. out-of-round hole　失圆孔
17. bellmouthed　钟形口，喇叭口
18. rotary-type bushing　旋转衬（钻）套

Reading Materials

Basic Operations of Machining Metal

There are hundreds of operations performed on metal parts by machine tools, but fundamentally they may be divided into the following five basic kinds：

1. Drilling. Drilling is the operation of producing holes in solid metal. A rotating drill called a twist drill is employed. Machine tools for drilling holes are called drilling machines or drill presses. There are many types and sizes of drilling machines. These machines can perform other operations besides drilling. The workpiece is held stationary, that is, clamped in position, and as the drill rotates, it is fed into the workpiece.

2. Turning and Boring. The engine lathe is the most common type of machine tool for

turning work. Turning is the operation of cutting or removing metal from a workpiece. A cutting tool is fed into or along the workpiece while the workpiece revolves. Boring is the operation of enlarging or machining a hole that has been drilled or cast into the metal. Boring on a lathe is done by feeding a single-point cutting tool into the workpiece as it revolves.

3. Milling. Milling is the operation of removing metal by means of a rotating cutting tool. The cutting tools have multiple cutting edges and are called milling cutters.

4. Grinding. Grinding is the operation of cutting or removing metal by means of an abrasive wheel called a grinding wheel. Grinding finishes work very accurately and smoothly. When grinding round work, the workpiece revolves as it is fed against the turning wheel. When grinding flat work, the workpiece is passed back and forth under the turning wheel.

5. Shaping, planing and slotting. These operations are done to produce accurate flat surfaces, using single-point cutting tools. We should understand the difference between the shaper, planer and slotting machine. On a shaper, the cutting tool moves back and forth over the work, while the work is fed against the tool. On a planer, the work moves back and forth under the cutting tool, while the cutting tool is fed into or against the workpiece. A slotting machine is really a vertical shaper, the cutting tool of which moves up and down. In this case, the work is fed against the cutting tool in a straight line or in a circle, depending upon the type of work being machined.

Reaming

Jig design for reaming is basically the same as for drilling, which has been discussed throughout the chapter. The main difference is the need to hold closer tolerances on the jigs and bushings, and provide additional support to guide the reamer. For long holes, it is essential to guide the reamer at both ends, as shown in Figure 20-2a, using special piloted reamers designed for this purpose. Jigs should be designed so that the pilot enters the bushing before the reamer enters the workpiece, and remains piloted until the reaming operation is completed. For short holes, the

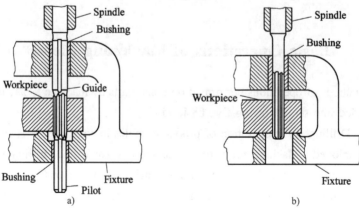

Figure 20-2   Fixtures for guiding reamers

reamer is usually guided at one end only, as seen in Figure 20-2b, with the bushing sized to fit the OD of the reamer. Additionally, bushings for reaming are generally longer than for drilling, usually three or four times the reamer diameter. Chip clearance is also generally less for reaming than for drilling, varying from one-fourth to one-half the tool diameter down to a maximum of 0. 125" ~0. 24" regardless of the reamer diameter.

Bushing bores must be closely controlled. Bushings that are too small can cause tool seizure and breakage. Bushings that are too large will result in bellmouthed or out-of-round holes. Data in the *Handbook of Jig and Fixture Design* can be used, as a guide.

Carbide bushings should be considered for long production runs or where abrasive conditions are present. Roller or ball bearing, rotary-type bushings also provide maximum wear while maintaining close tolerances under high loads.

Lesson 21　Fundamentals of Turning
and Boring on Lathe

Text

Turning is the process of machining external cylindrical and conical surfaces. It is usually performed on a lathe. Photographs and schematics of lathes are shown later in this chapter. As indicated in Figure 21-1, relatively simple work and tool movements are involved in turning a cylindrical surface. The workpiece is rotated into a *longitudinally* fed single-point-cutting tool. If the tool is fed at an angle to the axis of rotation, an external conical surface results. This is called taper turning. If the tool is fed at 90° to the axis of rotation, using a tool that is wider than the width of the cut, the operation is called facing, and a flat surface is produced[1].

By using a tool having a specific form or shape and feeding it *radial* or *inward* against the work, external cylindrical, conical, and irregular surfaces of limited length can also be turned. The shape of the resulting surface is determined by the shape and size of the cutting tool. Such machining is called form turning. If the tool is fed all the way to the axis of the workpiece, it will be cut in two. This is called parting or cutoff and a simple, thin tool is used. A similar tool is used for *necking* or partial cutoff.

Boring is a variation of turning. Essentially boring is internal turning. Boring can use single-point-cutting tools to produce internal *cylindrical* or *conical* surfaces. It does not create the hole but rather, machines or opens the hole up to a specific size. Boring also can be done on most machine tools that can do turning. However, boring also can be done using a rotating tool with the workpiece remaining stationary. Also, specialized machine tools have been developed that will do boring, drilling, and reaming, but will not do turning. Other operations, such as threading and knurling, can be done on machines used for turning. In addition, drilling, reaming, and tapping can be done on the rotation axis of the work.

Turning constitutes the majority of lathe work. The cutting forces, resulting from feeding the tool from right to left, should be directed toward the headstock to force the workpiece against the workholder and thus provide better work support[2].

If good finish and accurate size are desired, one or more roughing cuts usually are followed by one or more finishing cuts. Roughing cuts may be as proper chip thickness, tool life, lathe horsepower, and the workpiece permit. Large depths of cut and smaller feeds are preferred to the reverse procedure, because fewer cuts are required and less time is lost in reversing the carriage and resetting the tool for the following cut.

On workpieces that have a hard surface, such as casting or hot-rolled materials containing mill *scale*, the initial roughing cut should be deep enough to *penetrate* the hard materials. Otherwise, the entire cutting edge operates in hard, abrasive material throughout the cut, and the tool will dull rapidly. If the surface is unusually hard, the cutting speed on the first roughing cut should be reduced accordingly[3].

Finishing cuts are light, usually being less than 0.015in in depth, with the feed as fine as necessary to give the desired finish. Sometimes a special finishing tool is used, but often the same tool is used for both roughing and finishing cuts. In most cases one finishing cut is all that is required. However, where exceptional accuracy is required, two finishing cuts may be made. If the diameter is controlled manually, a short finishing cut (1/4in long) is made and the diameter checked before the cut is completed[4]. Because the previous micrometer measurements were made on a rougher surface, it may be necessary to reset the tool in order to have the final measurement, made on a smoother surface, check exactly.

In turning, the primary cutting motion is rotational with the tool feeding parallel to the axis of rotation (Figure 21-1). The rpm value of the rotating workpieces, N, establishes the cutting velocity V, at the cutting tool. The feed f_r is given in inches per revolution (in/rev). The depth of cut is d where:

$$d = \frac{D_1 - D_2}{2} \tag{21-1}$$

The length of cut is the distance traveled parallel to the axis, L plus some **allowance** or **overrun**, A to allow the tool to enter and/or exit the cut.

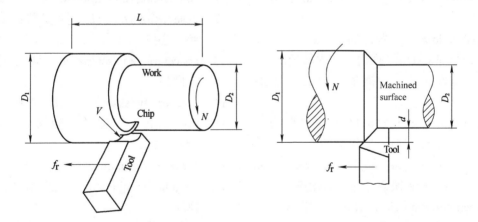

Figure 21-1　Basics of the turning process, normally performed in a lathe

Once the cutting speed, feed and depth of cut have been selected for a given material being cut with a tool of known cutting-tool material, the rpm value for the machine tool can be determined:

$$N = \frac{12V}{\pi D_1} \tag{21-2}$$

(using the larger diameters,) where the value 12 converts feet to inches.

Boring always involves the enlarging of an existing hole, which may have been made by drill or may be the result of a core in a casting[5]. An equally important and **concurrent pose** of boring may be to make the hole *concentric* with the axis of rotation of the workpiece and thus correct any *eccentricity* that may have resulted from the drill *drifting* off centerline. Concentricity is an important *attribute* of bored holes.

When boring is done in a lathe, the work usually is held in a chuck or on a faceplate. Holes may be bored straight, tapered, or to *irregular contours*. Figure 21-2 shows their relationship of the tool and the workpiece for boring. Boring is essentially internal cutting, while feeding the tool parallel to the rotation axis of the workpiece.

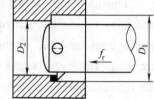

Figure 21-2　Boring

Questions

1. What does the term *turning* mean? What does the term *boring* mean?

2. Is it right to select small depth of cut and high cutting velocity? Why?

3. What do the terms *parting* or *cutoff* mean? What are the characteristics of the tool used in this machining?

4. What the function does the number 12 do in the equation (21-2)?

New Words and Expressions

1. longitudinally/lɔndʒiˈtjuːdinəli/*ad.* 轴向，纵向

2. radial/ˈreidiəl/*a.* 径向的

3. inward/ˈinwəd/*a.* 向内的，内部的

4. necking/nekiŋ/*n.* 颈缩，切颈，切退刀槽

5. cylindrical/siˈlindrik(ə)l/*a.* 圆柱面的

6. conical/ˈkɔnikəl/*a.* 圆锥面的

7. scale/skeil/*n.* 鳞，鳞片状物，刻度，秤盘　mill ~ 轧制铁鳞（轧制氧化皮）

8. penetrate/ˈpenitreit/*vt. & vi.* 穿入，切入，识破

9. allowance/əˈlauəns/*n.* 加工余量，公差

10. overrun/əuvəˈrʌn/*n.* 超越，超越行程

11. concurrent/kənˈkʌrənt/*a.* 同时，兼任

12. pose/pəuz/*vt. & vi.* 摆姿势；*n.* 姿势，姿态

13. concentric/kənˈsentrik/*a.* 同心的，同心圆的

14. eccentricity/eksenˈtrisiti/*n.* 古怪；偏心

15. drift/drift/*n.* 位移，偏移；*vt. & vi.* 飘动，漂流

16. attribute/ˈætribjuːt/*n.* 性质，属性，象征

17. irregular/iˈregjulə/*a.* 不规则的，非正规的

18. contour/ˈkɔntuə/*n.* 轮廓，外形

19. rpm = revolutions per minute　每分钟转数

Notes

[1] If the tool is fed at 90° to the axis of rotation, using a tool that is wider than the width of the cut, the operation is called facing, and a flat surface is produced.

如果车刀沿与旋转轴成90°角方向进给，所用的刀具宽度比要车削的宽度宽，就形成了一个平面，这种操作叫作端面车削。

本句的主句为 the operation is called facing, if 引导条件状语从句, and 引导结果从句, using 引导的分词短语又套了一个从句, 说明车刀特点。

[2] The cutting forces, resulting from feeding the tool from right to left, should be directed toward the headstock to force the workpiece against the workholder and thus provide better work support.

当车刀自右向左进给时将产生切削力。切削力应指向床头主轴箱, 以迫使工件紧靠在卡盘上, 这种方式有利于工件的夹持。

resulting from... 分词短语说明切削力产生的原因, to force... 不定式短语是宾语 headstock 的补足语, 进一步说明切削力指向主轴箱的结果。

[3] If the surface is unusually hard, the cutting speed on the first roughing cut should be reduced accordingly.

如果（材料）表面非常硬, 在第一次粗车时要相应地降低切削速度。

本句将 according to 这一常见形式变为 accordingly 形式放在句尾, 以强调第一次粗车时应降低速度。

[4] If the diameter is controlled manually, a short finishing cut (1/4in long) is made and the diameter checked before the cut is completed.

如果由人工控制直径, 则应在完成全部精车之前, 先车一小段（如1/4in 长）, 检查一下直径。

if 引导条件状语从句; before 引导时间状语从句, 以说明精车之前应检查直径; made 和 checked 并列, 与 is 构成被动态, 为主句 a short finishing cut 的谓语。

[5] Boring always involves the enlarging of an existing hole, which may have been made by drill or may be the result of a core in a casting.

镗削总是扩大已有的孔, 这个孔可能是钻出的, 也可能是铸造时由型芯形成的。

本句中的 involve 可译为"牵涉"或"涉及", 原句直译为"镗削总是牵涉（涉及）已经存在的孔的扩大"。

Glossary of Terms

Look the Figure 21-3 to learn these terms as following:

1. centre (Am. center) lathe　普通卧式车床
2. headstock with gear control (geared headstock)　主轴箱
3. reduction drive lever　降低驱动手柄
4. lever for normal and coarse threads　常规粗牙螺纹手柄
5. speed change lever　变速手柄
6. lead screw reverse-gear lever　丝杠反向齿轮手柄
7. change-gear box　变速箱
8. feed gear box (Norton tumbler gear)　进给箱
9. levers for changing the feed and thread pitch　进给和螺距调节手柄
10. feed gear lever (tumbler lever)　进给齿轮手柄
11. switch lever for right or left hand action of main spindle　主轴正反转操作手柄
12. lathe foot (footpiece)　支座

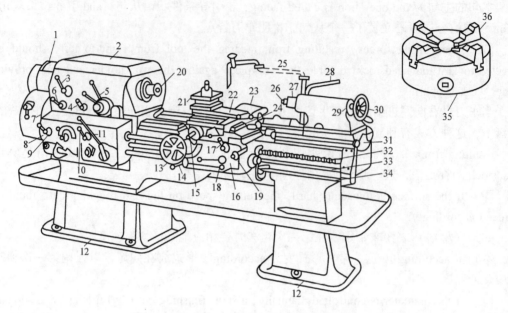

Figure 21-3 The lathe

13. lead screw handwheel for traversing of saddle (longitudinal movement of saddle) 溜板移动手轮

14. tumbler reverse lever 转向器反向手柄

15. feed screw 进给螺杆

16. apron (saddle apron, carriage apron) 溜板箱

17. lever for longitudinal and transverse motion （溜板自动）长运动手柄

18. drop (dropping) worm (feed trip, feed tripping device) for engaging feed mechanisms 进给啮合机构松开螺母

19. lever for engaging half nut of lead screw (lever for clasp nut engagement) 丝杠开合螺母手柄

20. lathe spindle 主轴

21. tool post 刀架

22. top slide (tool slide, tool rest) 小溜板

23. cross slide 中溜板

24. bed slide 溜板底板

25. coolant supply pipe 切削液管

26. tailstock centre (Am. center) 尾座顶尖

27. barrel (tailstock barrel) 套筒

28. tailstock barrel clamp lever 锁紧手柄

29. tailstock 尾座

30. tailstock barrel adjusting handwheel 尾座（套筒调节）手轮

31. lathe bed 床身

32. lead screw 丝杠

33. feed shaft 进给轴（光杠）

34. reverse shaft for right and left hand motion and engaging and disengaging 操纵杆（为左手和右手操作及进给齿轮啮合和脱离啮合的操纵轴）

35. four-jaw chuck (four-jaw independent chuck) 四爪卡盘

36. gripping jaw 颚形夹爪

Reading Materials

The Lathe

The lathe is one of the most useful and versatile machines in the workshop, and is capable of carrying out a wide variety of machining operations. The main components of the lathe are the headstock and tailstock at opposite ends of a bed, and a tool-post between them, which holds the cutting tool. The tool-post stands on a cross-slide which enables it to move sideward across the saddle or carriage as well as along it, depending on the kind of job it is doing.

The ordinary center lathe can accommodate only one tool at a time on the tool-post, but a turret lathe is capable of holding five or more tools on the revolving turret. The lathe bed must be very solid to prevent the machine from bending or twisting under stress.

The headstock incorporates the driving and gear mechanism, and a spindle which holds the workpiece and causes it to rotate at a speed which depends largely on the diameter of the workpiece. A bar of large diameter should naturally rotate more slowly than a very thin bar, the cutting speed of the tool is what matters. Tapered centres in the hollow nose of the spindle and of the tailstock hold the work firmly between them. A feed-shaft from the headstock drives the tool-post along the saddle, either forwards or backwards, at a fixed and uniform speed. This enables the operator to make accurate cuts and to give the work a good finish. Gears between the spindle and the feed-shaft control the speed of rotation of the shaft, and therefore the forward or backward movement of the tool-post. The gear, which the operator will select, depends on the type of metal, which he is cutting, and the amount of metal he has to cut off. For a deep or roughing cut the forward movement of the tool should be less than for a finishing cut.

Centers are not suitable for every job on the lathe. The operator can replace them by various types of chucks, which hold the work between jaws, or by a front-plate, depending on the shape of the work and the particular cutting operation. He will use a chuck, for example, to hold a short piece of work, or work for drilling, boring or screw-cutting. A transverse movement of the tool-post across the saddle enables the tool to cut across the face of the workpiece and give it a flat surface. For screw cutting, the operator engages the lead screw, a long screwed shaft which runs along in front of the bed and which rotates with the spindle. The lead screw drives the tool post, forwards along the carriage at the correct speed, and this ensures that the threads on the screw are of exactly the right pitch. The operator can select different gear speeds, and this will alter the ratio of spindle and lead screw speeds and therefore alter the pitch of the threads. A reversing lever on the headstock enables him to reverse the movement of the carriage and so bring the tool back to its original position.

Lathe Chuck

Lathe chuck consists of a body with inserted workholding jaws that slide radially in slots and are actuated by various mechanisms such as screws, scrolls, levers, and cams, alone and in a variety of combinations. The number of jaws varies. Chucks in which all the jaws move together are self centering and are used primarily for round work. Two-jaw chucks operate somewhat like a vise, and may be used for round and for irregular shaped workpieces by the use of suitably shaped jaws or jaw inserts.

The accuracy of a chuck deteriorates with usage owing to wear, dirt, and deformation caused by excessive tightening. Independent jaw chucks permit each jaw to move independently for chucking irregular-shaped workpieces or to center a round workpiece.

The jaws of most lathe chucks can be reversed to switch from external to internal chucking. Jaws may be adapted to fit workpiece shapes that are not round. The means of attaching a lathe chuck to different machine tools have been well standardized, so that chucks made by different manufacturers can be easily interchanged.

In addition to their standard jaws, lathe chucks may also be fitted with a variety of special purpose jaws to accommodate different types of workpiece surfaces and configurations. The principal types of chuck jaws used for these purposes are called soft jaws and are generally made of cast aluminum.

Lesson 22   Milling

Text

Milling is a basic machining process by which a surface is generated by **progressive** chip removal. The workpiece is fed into a rotating cutting tool. Sometimes the workpiece remains stationary, and the cutter is fed to the work. In nearly all cases, a multiple-tooth cutter is used so that the material removal rate is high. Often the desired surface is obtained in a single pass of the cutter or work, and because very good surface finish can be obtained, milling is particularly well suited and widely used for mass-production work. Many types of milling machines are used, ranging from relatively simple and **versatile** machines that are used for general-purpose machining in job shops and tool-and-die work on highly **specialized** machines for mass production[1]. Unquestionably, more flat surfaces are produced by milling than by any other machining process.

The cutting tool used in milling is known as a milling cutter. Equally spaced **peripheral** teeth will **intermittently engage** and machine the workpiece. This is called **interrupted** cutting.

Milling operations can be classified into two broad **categories** called peripheral milling and face milling. Each has many variations. In peripheral milling the surface is generated by teeth located on the periphery of the cutter body (Figure 22-1). The surface is parallel with the axis of rotation of the cutter. Both flat and formed surfaces can be produced by this method, the cross section on the resulting surface **corresponding** to the axial contour of the cutter[2]. This process, often called **slab** milling, is usually performed on horizontal-spindle milling machines. In slab milling, the tool rotates (mills) at some rpm value N while the work feeds past the tool at a table feed rate f_m in inches per minute.

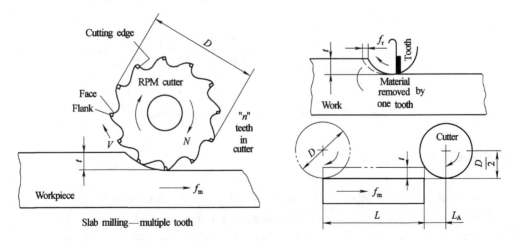

Figure 22-1   Basics of peripheral or slab milling process

As in the other processes, the cutting speed V and feed per tooth are "selected" by the engineer or the machine tool operator. As before, these variables depend on the work material, the tool material, and the specific process. The cutting velocity is that which *occurs* at the cutting edges of the teeth in the milling cutter[3]. The rpm of the spindle is determined from the surface cutting speed, where D is the diameter of cutter in inches, according to:

$$N = \frac{12V}{\pi D} \tag{22-1}$$

The depth of cut, called t in Figure 22-1, is simply the distance between the old and new machine surfaces. The width of cut is the width of the cutter or the work, in inches, and is given the symbol W. The length of the cut, L, is the length of the work plus some allowance L_A for approach and over travel. The feed rate of the table, f_m, in inches per minute is related to the amount of metal each tooth removes during a revolution (this is called the feed per tooth), f_1[4], according to:

$$f_m = f_1 Nn \tag{22-2}$$

where n is the number of teeth in the cutter (teeth/*rev*).

In face milling and end milling, the surface generated is at right angles to the cutter axis (Figure 22-2). Most of the cutting is done by the peripheral *portions* of the teeth, with the face portions providing some finishing action. Face milling is done on both horizontal and vertical-spindle machines.

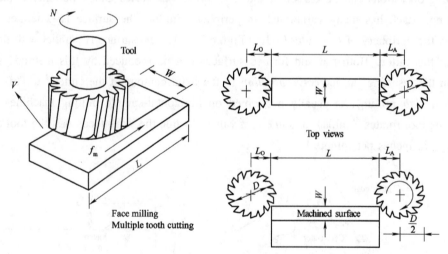

Figure 22-2   Basics of face and end milling process

Questions

1. What are the characteristics of milling comparing with other metal cutting?
2. Why is the milling particularly well suited and widely used for mass production?
3. What is the peripheral milling? And can use it on vertical-spindle machine?
4. What is face and end milling? Only a small part of the cutting is done by the peripheral portions of the teeth in face milling, true or false?

New Words and Expressions

1. progressive/prəˈgresiv/a.　逐步前进的，循序渐进的；n.　进步分子

2. versatile/ˈvəːsətail/a.　万能的，万象的，多变的

3. specialize/ˈspeʃəlaiz/v.　专门做，使专业化

4. peripheral/pəˈrifərəl/a.　周边的，圆周的；n.　外围设备

5. intermittent/intəˈmitənt/a.　间歇的，中断的，周期性的

6. engage/inˈgeidʒ/vt.　啮合，咬入，切入；从事，约定

7. interrupted/intəˈrʌptid/a.　中断的，被打断的，interrupted cutting 断续切削

8. category/ˈkætigəri/n.　种类，部门，类别

9. corresponding/kɔriˈspɔndiŋ/a.　相应的，符合一致的

10. slab/slæb/n.　（平）板，（厚）片，slab milling 面铣

11. occur/əˈkəː/v.　发生，被想到(to)，存在

12. rev/rev/n.　一次回转，旋转

13. portion/ˈpɔːʃən/n.　部分，区段

Notes

[1] Many types of milling machines are used, ranging from relatively simple and versatile machines that are used for general-purpose machining in job shops and tool-and-die work on highly specialized machines for mass production.

多种型号的铣床被广泛使用，范围包括在生产车间中用于常规加工的简易铣床和万能铣床，也包括在工模具车间使用的用于大批量生产的各种高度专业化铣床。

many 引导的短句是主句；分词 ranging 引导的短语作状语，进一步说明铣床的种类和用途，该分词短语中又有由 that 引出的定语从句，用来修饰 that 前面的 machines 一词。整个句子较长，翻译时只能意译。

[2] Both flat and formed surfaces can be produced by this method, the cross section on the resulting surface corresponding to the axial contour of the cutter.

用这种方法可以进行平面铣削和成形表面铣削，成形表面的横截面形状与铣刀的轴向轮廓一致。

句中 "," 后是一个分词独立结构，用作状语，表示结果，其逻辑主语为 the cross section（横截面），分词为 corresponding to（与……一致）。

[3] The cutting velocity is that which occurs at the cutting edges of the teeth in the milling cutter.

切削速度就是指铣刀齿切削刃上的速度。

that 引导表语从句；which 在表语从句中作主语。用此句型意在强调速度的定义。

[4] The feed rate of the table, f_m, in inches per minute is related to the amount of metal each tooth removes during a revolution (this is called the feed per tooth), f_1, ...

工作台的进给量与铣刀每一齿在铣刀每一转中切削的金属量（即每齿进给量）有关，进给量以 f_m 表示，单位为英寸/分钟，而每齿进给量以 f_1 表示，……

Glossary of Terms

Look the Figure 22-3 to find the English word groups according to the ideas of these Chinese words as follows:

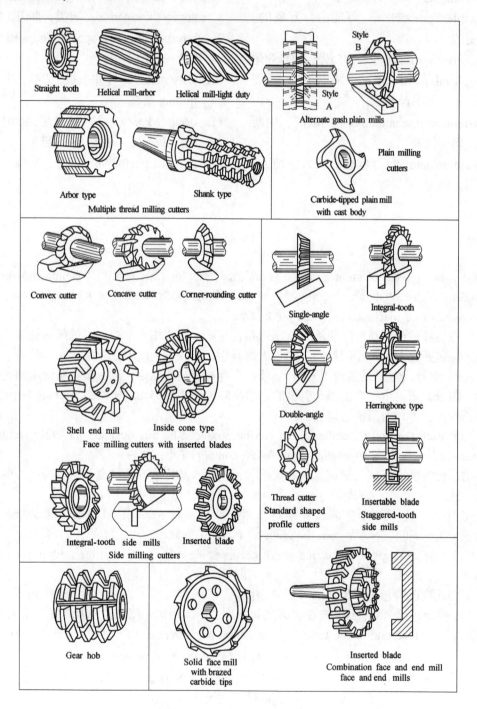

Figure 22-3   Common types of milling cutters

1. 标准螺旋角圆柱铣刀
2. 大螺旋角圆柱铣刀
3. 直齿铣刀
4. 单角铣刀
5. 双角铣刀
6. 镶齿套式面铣刀
7. 齿轮滚刀
8. 半凸圆成形铣刀
9. 凹半圆成形铣刀
10. 三面刃铣刀
11. 空心端铣刀
12. 错齿三面刃铣刀
13. 镶齿三面刃铣刀
14. 镶齿端-面组合铣刀
15. 多螺旋槽铣刀
16. 交错齿平铣刀
17. 凸圆角成形铣刀
18. 整体型铣刀
19. 盘式多螺旋槽铣刀
20. 铸铁刀体硬质合金平面铣刀

Reading Materials

Milling Cutters

Milling cutters are cylindrical cutting tools with cutting teeth spaced around the periphery (Figure 22-3). Another figure (abridged by editor) shows the basic tooth angles of a solid plain milling cutter. A workpiece is traversed under the cutter in such a manner that the feed of the workpiece is measured in a plane perpendicular to the cutter axis. The workpiece is plunged radially into the cutter, and sometimes, in rare cases, there is also an axial feed of the cutter, which results in a generated surface on the workpiece. Milling-cutter teeth intermittently engage the workpiece with the chip thickness being determined by the motion of the workpiece, the number of teeth in the cutter, the rotational speed of the cutter, the cutter lead angle and the overhang of the cutter on the workpiece.

There are two modes of operation for milling cutters. In conventional (up) milling the workpiece motion opposes the rotation of the cutter (Figure 22-4a), while in climb (down) milling the rotational and feed motions are in the same direction (Figure 22-4b). Climb milling is preferred wherever it can be used since it provides a more favorable metal-cutting action and generally yields a better surface finish. Climb milling requires more rigid equipment and there must be no looseness in the workpiece feeding mechanism since the cutter will tend to pull the workpiece.

Index milling cutters have precision ground carbide inserts positioned around the cutter body and are held in pins or wedges which can be released for indexing. Some milling cutters may have either profile-sharpened or form-relieved teeth. Profile-sharpened cutters are those, which are sharpened on the relief surface using a conventional cutter grinding machine. Form-relieved cutters are made with uniform radial relief behind the cutting edge. They are sharpened by grinding the face of the teeth. The profile style provides greater flexibility in adjusting relief angles for the job to be done, but it is necessary that any form on the cutter be reproduced during each resharpening. In the form-relieved style, the relief angle cannot be changed since it is fixed in the manufacture of the cutter. However, the form-relieved construction is well adapted to cutters with intricate profiles since the profile is not changed by resharpening.

Most large milling cutters are provided with an axial hole for mounting on an adapter or arbor, and usually have a drive key slot. Certain small-diameter cutters and some cutters for specialized applications are made using an integral shank construction where the cutting section is at the end of a straight or tapered shank which fits into the machine tool spindle or adapter. Also, some large facing cutters are designed to mount directly on the machine tool spindle nose.

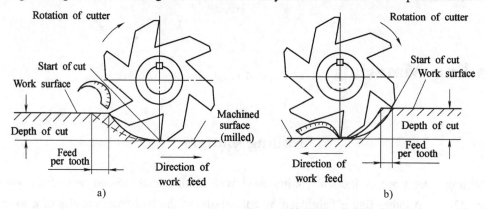

Figure 22-4　Modes of operation for milling cutters
a)　"Out-cut" "Conventional" or "Up" milling; also called "Against the cutter"
b)　"In-cut" or "Down" milling; also called "Feeding with the cutter"

Controls of Horizontal Milling Machines

The various controls of a typical horizontal milling machine are shown in Figure 22-5. These are identical to those of a horizontal machine.

Spindle speeds are selected through the levers 4, and the speeds indicated on the change dial 5. The speeds must not be changed while the machine is running. An 'inching', button 3 is situated below the gear change panel and, if depressed, 'inches' the spindle and enables the gears to slide into place when a speed change is being carried out. Alongside the 'inching' button is the switch for controlling the cutting-fluid pump 1 and one for controlling the direction of spindle rotation 2. The feed rates are selected by the lever 9 and are indicated on feed-rate dial.

To engage the longitudinal table feed, lever 8 is moved in the required direction—right for right feed, left for left feed. Adjustable trip dogs 6 are provided to disengage the feed movement at any point within the traverse range. Limit stops are incorporated to disengage all feed movements in the extreme position, to prevent damage to the machine in the event of a trip dog being missed.

To engage cross or vertical traverse lever 12 is moved up or down. The feed can then be

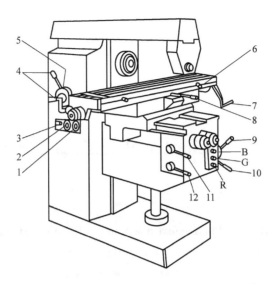

Figure 22-5   The horizontal milling machine

1—cutting-fluid pump   2—switch for controlling the direction of spindle rotation

3—button   4—spindle speed lever   5—speed change dial

6—trip dogs   7—single crank handle   8—feed lever   9—feed rate lever

10—feed direction lever   11—up-feed and down-feed lever   12—vertical traverse lever

engaged by moving lever 11 in the required direction. With cross traverse selected, movement of lever 11 upwards produces in-feed of the saddle, moving it downwards produce out-feed of the saddle.

With vertical traverse selected, movement of lever 11 upward produces up-feed to the knee, moving it downwards produces down-feed to the knee.

Rapid traverse in any of the above feed directions is engaged by an upward pull of lever 10. Rapid traverse continues as long as upward pressure is applied. When released, the lever will drop into the disengaged position. Alternative hand feed is provided by means of a single crank handle 7, which is engaged by slight pressure towards the machine. Spring ejectors disengage the handle on completion of the operation, for safety purposes, i. e. the handle will not fly round when feed or rapid traverse is engaged. The single crank handle is interchangeable on table, saddle, and knee movements.

Unit Eight

Lesson 23   Electrochemical Machines

Text

The principle *underlying electrochemical* machining (ECM) rests on an exchange of charge and material between a positively charged anodic workpiece and a negatively charged *cathode* tool in an *electrolyte*. In such conditions, the *anode* dissolves *whilst* the cathode (tool) is not affected. The volume of metal removal may be calculated according to Faraday's Law:

$$V = CIt$$

Where   C—a constant dependent on work material;

I—the current flowing between the tool and the work;

t—the time of *erosion*.

The current is dependent on the *gap* between the tool and the work, the area of erosion and the conductivity of the electrolyte, as well as the supply voltage. The working gap maintained between the *electrode* and work allows machining to take place without physical contact[1]. The electrolyte (e. g. NaCl or $NaNO_3$ in water solution) is pumped into the working gap and also serves as a *coolant* which is necessary due to the high energy density. The material which has been eroded from the work forms a *sludge*, and must be separated from the electrolyte by *filter* or *centrifuges*.

Electrochemical Die-sinking Machines. Figure 23-1 presents, in *schematic* form, the main components of an electrochemical die-sinking machine. A feeding device advances the tool

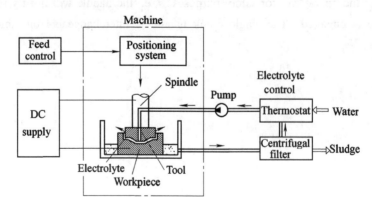

Figure 23-1   Main components of an electrochemical machining plant

towards the work in accordance with the rate of metal removal. When producing internal forms, a design problem arises in relation to the shape and size of the tool. The gap is not constant, but is a function of the state of the surface to be eroded and the rate of tool advance. If, for example, a *cylindrical* bore is to be sunk, a simple cylindrical tool (as shown in Figure 23-2) is not suitable. This would result in a constantly increasing gap size and the current density would decrease in proportion (Figure 23-2 left). With a tool suitably insulated on its sides (Figure 23-2 right), the *offending* excessive erosion of cylindrical sides will be suppressed.

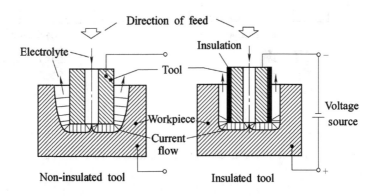

Figure 23-2　Cavity development during electrochemical
machining with different tool electrodes

The functions which must be performed by the individual elements of the machine are *summarized* in table. Owing to the small gap sizes which are used in electrochemical machining, high electrolyte pressures (>20 bar) are necessary so that there is an adequate flow for effective cooling and removal of the eroded material. In an erosion area of 10000 mm^2, forces in excess of 20kN may be experienced, with which the tool-feeding system and the machine structure must be able to cope. As the electrolyte consists of a corrosive salt solution, all machine components likely to come into contact with it must be corrosion-proof. An important *auxiliary* installation is a short-circuit cut-out, which immediately stops the supply of further electrical energy in the event of inadequate clearance of the eroded material and insufficient gap between tool and work.

Electrochemical die-sinking machines are constructed in an open 'C' structure for small to medium sized work, and for machines having larger work areas the closed 'O' construction form is used.

The preferred direction of tool feed is vertical. Figure 23-3 shows a machine built on a building-block system, in which various combinations of machine bed, frame elements and work heads may be achieved. The tool advance is mostly electrohydraulic; however, electric drives such as *disc armature* motors connected to a *recirculating* ball lead screw and nut are also used[2].

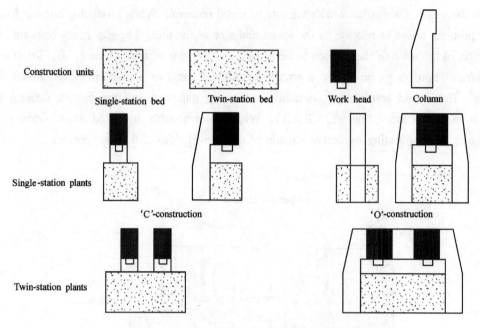

Construction units

Single-station bed   Twin-station bed   Work head   Column

Single-station plants

'C'-construction   'O'-construction

Twin-station plants

Figure 23-3   Unit construction of electrochemical machining plants

Questions

1. What is the Faraday's Law?
2. Describe the principle of electrochemical machining.
3. What are elements influencing the current in Faraday's Law?

New Words and Expressions

1. underlying /ˌʌndəˈlaiiŋ/ a. 在下的，基础的，潜在的

2. electrochemical/iˈlektrəuˈkemikəl/ a. 电化学的

3. cathode/ˈkæθəud/ n. 阴极，负极

4. electrolyte/iˈlektrəulait/ n. 电解质，电解液，电离质

5. anode/ˈænəud/ n. 阳极，正极(anodic a.)

6. whilst/hwailst/ conj. = while 同时，有时

7. erosion/iˈrəuʒən/ n. 腐蚀，侵蚀，磨蚀，冲蚀

8. gap/gæp/ n. 裂口，缺口，间隙，火花隙

9. electrode/iˈlektrəud/ n. 电极

10. coolant /ˈkuːlənt/ n. 冷却液，切削液，冷却剂

11. sludge/slʌdʒ/ n. 淤泥，泥浆

12. filter/ˈfiltə/ n. 滤器，滤纸；vt. 过滤

13. centrifuge/ˈsentrifjudʒ/ n. 离心机

14. schematic/skiˈmætik/ a. 图解的，纲要的，按照图示的

15. cylindrical/siˈlindrikəl/ a. 圆柱体的，圆柱形的，圆筒状的

16. offend/əˈfend/ vt. 冒犯，使不愉快；vi. 违反

17. summarize/ˈsʌməraiz/ n. 概括，概述，总结

18. auxiliary/ɔːgˈziliəri/ a. 辅助的，补助

的，附属的；*n.* 辅助者，附件

19. disc／disk／*n.* 圆盘，圆板，圆面，盘状物

20. armature／ˈɑːmətjuə／*n.* 转子，附件，加强料，铠装

21. recirculate／ˈriːˈsəːkjuleit／*v.* 再循环，重

复循环，回流

22. due to 由于，应归于

23. take place 发生

24. in accordance with 按照，依据，与……一致

25. in relation to 关于，涉及，与……有关

Notes

[1] The working gap maintained between the electrode and work allows machining to take place without physical contact.

电极和工件之间保持的工作间隙，既避免了物理接触，又能确保发生电解。

单个分词 working 作 gap 的前置定语，分词短语 maintained between the electrode and work 作 gap 的后置定语。注意：当分词作定语时，一般情况下，单个分词位于被修饰词前，而分词短语则一定要位于被修饰词之后。

[2] The tool advance is mostly electrohydraulic; however, electric drives such as disc armature motors connected to a recirculating ball lead screw and nut are also used.

大多数情况下工具电极进给采用电液驱动，但也可以由电装置实现，例如，可以把伺服电动机连接到一个带有螺母的滚珠丝杠上。

在"；"后面的句子中，electric drives 是主语，而谓语 are also used 在句尾，二者相隔甚远，阅读时应该注意这一现象。

Glossary of Terms

1. underlying metal 底层金属
2. electroforming 电成形，电铸
3. perforated electrode 多孔电极
4. electro-chemical machining 电化学加工
5. form electromachining 电加工成形
6. electric machining 电加工
7. salt bath electrode furnace 电极盐浴炉
8. electrolytic forming 电解成形
9. electrolytic forming machine 电解成形机
10. electrochemical machining 电解加工
11. electrochemical machining tool 电解加工机床
12. electrolytic universal tool and cutter grinder 电解万能工具磨床
13. electrolytic heat treatment 电解液热

处理
14. electrohydraulic forming 电液成形
15. electrolytic deburring 电解去毛刺
16. electrolytic marking machine 电解刻印机
17. electrolytic surface grinder 电解平面磨床
18. chemical machining (CM) 化学加工
19. electrochemical grinding (ECG) 电化学磨削加工
20. laser-beam machining (LBM) 激光束加工
21. electron-beam machining (EBM) 电子束加工
22. chemical milling 化学铣削

Reading Materials

Chemical Etching Machines

In contrast to the electrochemical processes, the chemical etching process does not use a forming tool nor an external electric power. The work material is mainly removed as a result of differences of potential at the grain boundaries according to the particular material being worked. A variety of acids are used as activators. Filters or centrifuges separate the removed material from the etching medium.

The work is carried out either in a bath of the etching medium (dip etching) or by spraying the etching medium onto the work (spray etching). In order to obtain a particular work geometry, the parts of the surface which are not to be machined are protected by masking. Frequently, a photographic film technique is employed, whereby the workpiece is covered with a light-sensitive film by rolling it on or dipping. A photographic image of the desired work geometry is then projected onto the work surface, so that the illuminated areas of the work become sensitive to the acid attack and the remaining areas are suitably masked.

Ultrasonic Machining Installations

Ultrasonic machining installations are mainly used for machining electrically non-conductive, brittle materials (such as glass, ceramic oxides, precious stones, carbides, germanium, silicon, graphite and hard metals). A high frequency generator activates the magnetostrictive oscillator, which transmits the high-frequency oscillations to the tool soldered to the tapered bronze transformer. The tool itself is only indirectly active in the actual metal removal process. The work material is removed through abrasive grains suspended in a slurry, which acts in a manner similar to that of the lapping process—like a number of simultaneously acting chisel points. The slurry suspension is externally applied to the work area and sucked up through the transformer.

In many installations, no separate feeding mechanism is provided. By a vertical arrangement of the tool-work system, the tool advances into the work as a result of its own weight.

Lesson 24　Electrodischarge Machines

Text

When applying the electrodischarge machining （EDM） process, the material is eroded as a result of an electrical *discharge* between tool and work. Due to the resultant short-lived, but very high, temperature rises, metal particles at the point of discharge are molten, partially *vaporized* and removed from the melt by mechanical and electromagnetic forces. The working medium is a *dielectric*, which washes the eroded material away and simultaneously acts as a coolant.

As in the case of ECM, EDM is a copying process where there is no contact between tool and work. *Contrary* to ECM, however, there is some erosion of the tool in EDM, which must be allowed for in the tool design to ensure accuracy of machining.

A further difference arises from the fact that in EDM there is no fixed tool feed, but the gap size must be maintained in accordance with the rate of metal removal and the conditions existing within the gap[1].

Electrodischarge Die-sinking Machines. The construction principles of an EDM die-sinking machine are shown in Figure 24-1. Spark erosion takes place in a container filled with the dielectric, in which the work is clamped. The controlled feed of the electrode is through an electrohydraulic or electromechanical *servo* system. The electrical energy for erosion is provided by the erosion generator. The filtering unit separates the *eroded* material from the dielectric. In the upper left of Figure 24-1, a single discharge is illustrated in enlarged form. The applied voltage *ionizes* the gap at the beginning of the discharge. At the point of highest field strength, a channel is formed, through which the discharge current flows. At each end of the channel, the material melts and the channel and its *surrounding* gas bubble expand. When the voltage is fully discharged, the channel *collapses* and the molten material vaporizes, simulating a *miniature*

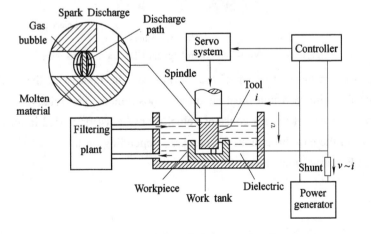

Figure 24-1　Electrodischarge erosion plant

explosion. The ***resultant crater*** is a ***hallmark*** of the irregular and ***scarred*** surface finish of spark-eroded work.

Electrodischarge Cutting Machines. An important application of the spark-erosion process is the cutting of metal by wire electrodes. The process is used for the production of ***apertures*** in cutting tools and the manufacture of tool electrodes for EDM. Figure 24-2 illustrates the principle. The cutting tool is a thin copper or brass wire, which enters the work during cutting without physical contact. This ***suffers*** wear as a result of the action of spark erosion, and for this reason fresh wire is constantly supplied. The ***apparatus*** required for wire feeding can be seen in Figure 24-3. The degree

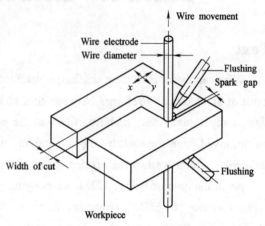

Figure 24-2   Numerically controlled EDM cutting x and y right-angled coordinates

of wire tension, the rate of wire consumption and the reach of the wire support arms are adjusted in accordance with the work to be done and the size of the workpiece[2]. The working medium for electrodischarge cutting is usually de-ionized water, which is fed to the work area with the use of ***flushing*** jets.

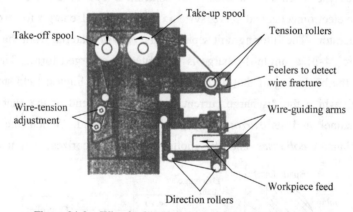

Figure 24-3   Wire-feed control for an EDM cutting machine

According to the required contour of the workpiece, the table with the work clamped to it and the slide with the wire feed unit must be suitably positioned. The relative advance of the cutting tool to the work does not have a constant velocity, but must be varied in accordance with the conditions existing in the gap throughout the process, depending on the progress of the cut, as was the case in electrodischarge machining.

Questions

1.   What are the differences between ECM and EDM?

2. List one of the important applications of the spark-erosion process.

3. Describe the operational principles of an electrodischarge die-sinking machine.

New Words and Expressions

1. discharge/disˈtʃɑːdʒ/v. 放出，排出，释放；n. 放电，流出，卸货
2. vaporize/ˈveipəraiz/v. （使）汽化，（使）蒸发
3. dielectric/daiiˈlektrik/n. 电介质，绝缘材料；a. 电介质的，绝缘的
4. contrary/ˈkɔntrəri/a. 相反的，逆行的，矛盾的
5. servo/ˈsəːvəu/n. 伺服，伺服机构
6. erode/iˈrəud/vt. vi. 腐蚀，侵蚀
7. ionize/ˈaiənaiz/vt. 使电离，离子化
8. surrounding/səˈraundiŋ/a. 周围的；n. 环境，外界
9. collapse/kəˈlæps/v. 崩溃，倒塌，破裂
10. miniature/ˈminiətʃə/n. 缩样，雏形；a. 小型的，微型的；v. 使成小型

11. explosion/iksˈpləuʒən/n. 爆发，爆炸，迅速增长
12. resultant/riˈzʌltənt/a. 作为结果而发生的，合成的；n. 结果，组合，合力
13. crater/kreitə/n. 陷口
14. hallmark/hɔːlˈmɑːk/n. 检验烙印，品质证明；vt. 在……上盖上纯度检验印记
15. scar/skɑː/n. 伤疤，斑疤；vt. 使留下伤痕
16. aperture/ˈæpətjuə/n. 隙，缝
17. suffer/ˈsʌfə/vt. 遭受，经受
18. apparatus/æpəˈreitəs/n. 机械，设备，装置，仪器
19. flushing/flʌʃiŋ/n. 冲洗，净化
20. in accordance with 与……一致，依照

Notes

[1] A further difference arises from the fact that in EDM there is no fixed tool feed, but the gap size must be maintained in accordance with the rate of metal removal and the conditions existing within the gap.

电火花加工与电解加工之间还有更重要的区别，事实上，电火花工具的加工进给量不是固定的，加工间隙的大小是依据去除金属的速率与该间隙的加工状态来确定的。

由 that 引出的从句是抽象名词 fact 的同位语从句，它实际上是由两个句子组成，中间用 "，" 隔开。句尾的分词短语 existing within the gap 是 the conditions 的后置定语，它们合起来可译为 "加工间隙的加工状态"。

[2] The degree of wire tension, the rate of wire consumption and the reach of the wire support arms are adjusted in accordance with the work to be done and the size of the workpiece.

依据加工要求和工件尺寸来调整金属丝的张紧程度、损耗率以及金属丝支撑臂所能达到的范围。

句中三个并列主语 the degree of..., the rate of..., the reach of... 属于 "the + 具有动作意义的名词 + of + 名词" 的结构，最后一个名词多数是行为对象。

Glossary of Terms

1. spark-erosion machining　电火花腐蚀加工
2. electrical discharge machining（EDM）电火花加工
3. servo system　伺服系统
4. electrodischarge cutting machine　电火花切割机
5. electrical discharge machine　电火花加工机床
6. electrical spark-erosion perforation　电火花腐蚀打孔
7. electrode contact surface　电极接触面
8. electrical discharge forming　电火花成形
9. laser cutting machine　激光切割机
10. electron beam cutting machine　电子束切割机
11. cavity sinking EDM machines　型腔电火花加工机床
12. EDM grinders　电火花加工磨床
13. travelling-wire EDM machine　线电极电火花加工机床
14. electro-discharge machine tool　电火花加工机床
15. electron beam machining（EBM）电子束加工
16. electron beam machine tool　电子束加工机床
17. form electromachining　电加工成形面
18. tiny hole spark-erosion grinding machine　电火花腐蚀小孔磨床
19. spark-erosion cutting with a wire　电火花腐蚀线切割
20. wire cut electric discharge machine　电火花线切割机

Reading Materials

Electron–Beam Cutting Installations

In electron-beam cutting, the work material is vaporized as a result of a beam of accelerated electrons impinging on a point of contact. Within a few milliseconds, a channel is cut into the work material. The vapour pressure forces the molten metal in the immediate vicinity out of the channel. The depth diameter and form of the cut can be controlled through the characteristics of the beam.

Apart from the actual work chamber, a high-voltage source (up to 150kV) is required, as are devices for the positioning of the electron beam and the work in relation to it. The process takes place in a vacuum in order to avoid the energy-absorbing collision of the electrons with air molecules. The beam may be deflected sideways and focused through the use of a system of magnetic lenses and deflection coils. The power density may be up to 10^6 kW/cm^2 with a minimum beam diameter of $2\mu m$.

Laser Cutting Machines

Laser cutting uses the erosion effect of high-energy light beams. As in the case of electron-beam cutting, the work material is vaporized at the point of impact. According to the laser material used, differentiation is made between solid and gas lasers. With solid lasers (e. g. ruby, neodymium-yttrium-aluminium-garnet), the excitement of a light emission is achieved with the use of a flash light (pump light), and when using a gas laser (e. g. CO_2, He-Ne) through the provision of a high voltage. A lens system focuses the monochromatic high-energy light. The power density achievable at the point of contact with the work may be up to 10^7 kW/cm^2.

General Characteristics of Electromachining Process

General characteristics of electromachining process are listed in the following table (Table 24-1).

Table 24-1 General characteristics of electromachining process

Process	Characteristics	Process parameters, typical material removal rate, or cutting speed
Electrochemical machining (ECM)	Complex shapes with deep cavities; highest rate of material removal among electromachining; expensive tooling and equipment; high power consumption; medium to high production quantity	5 ~ 25 DC 1. 5 ~ 8 A/mm^2 1. 2 ~ 2. 5 mm/min (depend on current density)
Electrochemical grinding (ECG)	Cutting off and sharpening hard materials, such as tungsten-carbide tools; also used as a honing process; higher removal rate than grinding	1 ~ 3 A/mm^2 typically 25 mm^2/s per 1000 A
Electrodischarge machining (EDM)	Shaping and cutting complex parts made of hard materials; some surface damage may result; also used as a grinding and cutting process; expensive tooling and equipment	50 ~ 380 V 0. 1 ~ 500 A; typically 300 mm^2/min
Wire EDM	Contour cutting of flat or curved surfaces; expensive equipment	Varies with material and thickness
Electron-beam machining (EBM)	Cutting and hole making on thin materials; very small holes and slots; heat-affected zone; requires a vacuum; expensive equipment	1 ~ 2 m/min
Laser-beam machining (LBM)	Cutting and hole making on thin materials; heat-affected zone; does not require a vacuum; expensive equipment; consumes much energy	0. 5 ~ 7. 5 m/min

Unit Nine

Lesson 25　Classification of Numerical Control Machines

Text

Numerical control machines can be generally classified using the following categories:

1. Power drives: *hydraulic*, *pneumatic* and electric.
2. Machine tool control: point-to-point and contouring (or continuous path).
3. Positioning systems: incremental positioning and *absolute* positioning.
4. Control loops: open-loop and closed-loop.
5. Coordinate definition: right-hand coordinate system and left-hand coordinate system.

Power Drives. One of the most notable *distinguishing* features of an industrial-quality NC machine is its power source. The power source usually determines the range of the machine's performance capabilities and in turns its *feasibility* for various applications. The three principal power sources are hydraulic, pneumatic, and electric.

Most high-power machines are generally driven by hydraulic power. Hydraulics can deliver large forces, so that the machine slides move with more *uniform* speed. *Offsetting* this advantage is cost, which is usually higher for hydraulic machines than for electric or pneumatic models of *equivalent* rating. Hydraulic power also requires additional peripherals, such as a *reservoir*, valves, etc. The major disadvantages of hydraulic power are the noise normally associated with these units, and hydraulic *contamination* from leaking fluid.

Pneumatic machines are often the least expensive power *alternative*. The availability of shop air at about 90 to 100 psi that can be *tapped* to power a pneumatic machine is an added advantage. Each axis on a pneumatic machine is normally controlled only at the end points. However, by varying the timing and *sequence*, an infinite *variation* of programmed setups is possible. The motion in a pneumatic machine is usually nonuniform in nature (typically called "bang-bang", with high acceleration and deceleration) [1].

Electric drives are the most applicable for precision jobs or when close precision control is desired. *Sophisticated* motion-control features are typical of electrical machines. There are two major groups among electric drives. One type uses stepper motors, which are driven a precise angular rotation for every voltage pulse issued by the controller [2]. Stepper motor movements can be very precise provided the *torque* load does not exceed the motor's design limits.

Because of this inherent accuracy, stepper motor systems are usually of the open-loop type.

The other kind of electric drive is the servo drive. These motors invariably *incorporate feedback* loops for the driven components back from the driver to a controller. There is continuous monitoring of positions and error conditions are promptly corrected by issuing appropriate voltage or current response changes until the position and velocity error goes to zero[3]. The servo drive is a continuous-position device whose position is measured by an encoded transducer, inductosyn, resolver, or other similar feedback device. Servo motors typically provide a smoother and more continuously controllable movement.

Motion Control. Point-to-point (PTP) control systems position the tool from one point to another within a coordinate system. Each tool axis is controlled separately, and the motion of the axes is either one axis at a time (Figure 25-1a) or multiple-axes motion (Figure 25-1b) with constant velocity on each axis. The control of motion is always defined by the programmed points, not by the path between them. The simplest example of such a system is a drilling machine, in which the workpiece or tool is moved along two axes until the center of the tool is positioned over the desired hole location. The drill path and its feed while traveling from one point to the next point are assumed to be unimportant[4]. The path from the starting point to the final position is not controlled. The data for the desired position is given by coordinate values. Rapid traverse is usually a point-to-point operation, even on contouring systems.

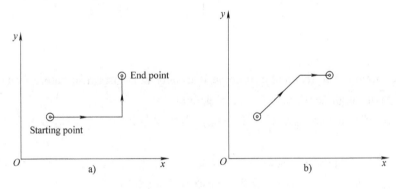

Figure 25-1　Point-to-point control
a) One axis at a time　b) Multiple-axes motion

Questions

1. What are the major advantages and disadvantages of hydraulic power for a NC machine?
2. What is an added advantage for the NC machine driven by pneumatic power?
3. What is the inherent advantage of stepper motor systems in NC machine?
4. The control of motion in PTP control system is that always defined both by programming points and by path between them, is this true? Why?

New Words and Expressions

1. hydraulic/hai'drɔ:lik/*a.* 水力的，液压的
2. pnuematic/nju:'mætik/*a.* 气动的，压缩空气推动的
3. absolute/'æbsəlu:t/*a.* 绝对的，无条件的，独立的
4. distinguish/dis'tiŋgwiʃ/*v.* 区分，分辨，显示……特性
5. feasibility/'fi:zəbiliti/*n.* 可行性，可做到
6. uniform/'ju:nifɔ:m/*a.* 不变的，一贯的；*n.* 制服
7. offset/'ɔfset/*vt.* 抵消，补偿
8. equivalent/i'kwivələnt/*a.* 相等的，相同的
9. reservoir/'rezəvwɑ:/*n.* 水库，油槽，油池
10. contamination/kəntæmi'neiʃən/*n.* 污染，污染物
11. alternative/ɔ:l'tə:nətiv/*a.* 可替代的，可选择的
12. tap/tæp/*n.* 水龙头，塞子；*v.* 开、放(水、气)等
13. sequence/'si:kwəns/*n.* 顺序，序列
14. variation/vɛəri'eiʃən/*n.* 变化，误差，偏差
15. sophisticate/sə'fistikeit/*v.* 改进，使完善，掺杂；曲解
16. torque/tɔ:k/*n.* 转矩，扭矩
17. incorporate/in'kɔ:pəreit/*v.* 使结合，含有，安装有
18. feedback/'fi:dbæk/*n.* 反馈

Notes

[1] The motion in a pneumatic machine is usually nonuniform in nature (typically called "bang-bang", with high acceleration and deceleration).

气动驱动的特点是机床运动通常不平稳，典型现象是在高速加速或减速时发出"砰砰"的响声。

本句虽然是一个简单句，但要很好地理解 nature 和 called 两个单词，前者译为"性质"，后者译为"发出……声音"，千万不能译为"叫作……"。

[2] One type uses stepper motors, which are driven a precise angular rotation for every voltage pulse issued by the controller.

一种类型是步进电动机。步进电动机能在控制器输出的每一个电压脉冲控制下精确转动一个角度。

which 引导的从句是宾语补足语，说明步进电动机的特性，由 are 可知 which 代表 motors，在从句中作主语。

[3] There is continuous monitoring of positions and error conditions are promptly corrected by issuing appropriate voltage or current response changes until the position and velocity error goes to zero.

不间断地监视运动位置，并通过提供与位置变化相适应的电压或电流信号迅速纠正偏差，直到位置和速度偏差降为零。

Glossary of Terms

1. open（close）loop　开（闭）环
2. servo motor（drive）　伺服电动机（传动）
3. stepper motor　步进电动机
4. synchro drive（system）　同步驱动（系统）
5. encoded transducer　编码传感器
6. inductosyn　感应同步器
7. resolver　解析器
8. compensator　补偿器
9. incremental measuring system　增量测量系统
10. analog control　模拟控制
11. assembly language　汇编语言
12. data processing system　数据处理系统
13. graphic data processing　图形数据处理
14. linear（circle）interpolator　线性（圆形）插补器
15. pounds per square inch（psi）　磅每平方英寸
16. Direct Numerical Control（DNC）　直接数字控制
17. Computer Numerical Control（CNC）　计算机数字控制
18. Data Processing Unit（DPU）　数据处理单元
19. Data Loops Unit（DLU）　数据循环单元
20. Control Loops Unit（CLU）　控制循环单元

Reading Materials

ISO Standards for Coding

Manual Part Programming Methods. In the earlier days, a number of formats for NC part programs were used such as fixed sequence or tab sequential. These systems required giving a large number of unwanted or duplicate information in each block of a part program. These have now been replaced by means of a system called Word Address Format in which each of the information or data to be input in the form of numerical digits is preceded by a word address in the form of an English alphabet. For example, N105 means that N is the address for the numerical data 105. Thus the controller can very easily and quickly process all the data entered in this format. A typical block of word address format may be looked as follows:

N115 G81 X20. 5 Y55. 0 Z-12. 0 R2. 0 F150 M3

ISO Standards for Coding. In the early years of development of Numerical Control, primary importance was given to standardization. Thus many of the things that we use in NC are standardized and many of manufacturers follow the standards to a great extent. One of the first things to be standardized was the word address used in programming. All the 26 letters of the English alphabet were standardized and given meaning shown in Table 25-1.

Table 25-1   Standardized word addresses used for CNC part programming

Character	Address for	Character	Address for
A	Angular dimension around x-axis	N	Sequence number
B	Angular dimension around y-axis	O	Reference rewind stop
C	Angular dimension around z-axis	P	Third rapid traverse dimension or tertiary motion dimension parallel to x[1]
D	Angular dimension around special axis or third feed function	Q	Second rapid traverse dimension or tertiary motion dimension parallel to y[1]
E	Angular dimension around special axis or second feed function	R	First rapid traverse dimension or tertiary motion dimension parallel to z[1]
F	Feed function	S	Spindle speed function
G	Preparatory function	T	Tool function
H	Unassigned	U	Secondary motion dimension parallel to x[1]
I	Distance to arc centre or thread lead parallel to x	V	Secondary motion dimension parallel to y[1]
J	Distance to arc centre or thread lead parallel to y	W	Secondary motion dimension parallel to z[1]
K	Distance to arc centre or thread lead parallel to z	X	Primary x motion dimension
L	Do not use	Y	Primary y motion dimension
M	Miscellaneous function	Z	Primary z motion dimension

[1] Where D, E, P, Q, R, U, V and W are not used as indicated, they may be used elsewhere.

Standardized Codes

ISO has standardized a number of these preparatory function also popularly called G codes. The standardized codes are shown below in Table 25-2.

Table 25-2   Standardized Codes

Code	Function	Code	Function
G00	Point-to-point positioning, rapid traverse	G41	Cutter radius compensation-offset left
G01	Line interpolation	G42	Cutter radius compensation-offset right
G02	Circular interpolation, clockwise (CW)	G43	Cutter compensation-positive
G03	Circular interpolation, anti-clockwise (CCW)	G44	Cutter compensation-negative
G04	Dwell	G45 ~ G52	Unassigned
G05	Hold/Delay	G53	Deletion of zero offset
G06	Parabolic interpolation	G54 ~ G59	Datum point/zero shift
G07	Unassigned	G60	Target value, positioning tolerance 1
G08	Acceleration of feed rate	G61	Target value, positioning tolerance 2, or loop cycle
G09	Deceleration of feed rate	G62	Rapid traverse positioning
G10	Linear interpolation for "long dimensions" (10 ~ 100in)	G63	Tapping cycle
G11	Linear interpolation for "short dimensions" (up to 10in)	G64	Change in feed rate or speed
G12	Unassigned	G65 ~ G69	Unassigned
G13 ~ G16	Axis designation	G70	Dimensioning in inch units
G17	*xy* plane designation	G71	Dimensioning in metric units
G18	*zx* plane designation	G72 ~ G79	Unassigned
G19	*yz* plane designation	G80	Canned cycle cancelled
G20	Circular interpolation, CW for "long dimensions"	G81 ~ G89	Canned drilling and boring cycles
G21	Circular interpolation, CW for "short dimensions"	G90	Specifies absolute input dimensions
G22 ~ G29	Unassigned	G91	Specifies incremental input dimensions
G30	Circular interpolation, CCW for "long dimensions"	G92	Programmed reference point shift
G31	Circular interpolation, CCW for "short dimensions"	G93	Unassigned
G32	Unassigned	G94	Feed rate/min (inch units when combined with G70)
G33	Thread cutting, constant lead	G95	Feed rate/rev (metric units when combined with G71)
G34	Thread cutting, linearly increasing lead	G96	Spindle feed rate for constant surface feed
G35	Thread cutting, linearly decreasing lead	G97	Spindle speed in revolutions per minute
G36 ~ G39	Unassigned	G98 ~ G99	Unassigned
G40	Cutter compensation-cancels to zero		

Note: Many of control manufacturers follow these standardized codes without altering the meaning. However, some manufacturers do change them to suit their way of programming.

Lesson 26　Machining Centers

Text

Types of Machining Centers. There are two main types of machining centers: the horizontal spindle machine center and the vertical spindle machine center.

Horizontal Spindle Type.

1. The traveling-column type is equipped with one or usually two tables on which the work can be mounted. With this type of machining center, the workpiece can be machined while the operator is loading a new workpiece on the other table.

2. The fixed-column type is equipped with a *pallet shuttle*. The pallet is a removable table on which the workpiece is mounted. After the workpiece has been machined, the workpiece and pallet are moved to a shuttle which then *rotates*, bringing a new pallet and workpiece into position for machining[1].

Vertical Spindle Type. The vertical spindle machining center is a *saddle*-type construction with sliding *bedways* which utilizes a sliding vertical head instead of a *quill* movement.

Parts of the CNC Machining Centers. The main parts of CNC machining centers are the bed, saddle, column, table, servo motors, ball screws, spindle, tool changer, and the machine control unit.

Bed. The bed is usually made of high-quality cast iron which provides for a rigid machine capable of performing heavy-duty machining and maintaining high precision. Hardened and ground ways are mounted to the bed to provide rigid support for all linear axes.

Saddle. The saddle, which is mounted on the hardened and ground bedways, provides the machining center with the x-axis linear movement.

Column. The column, which is mounted to the saddle, is designed with high *torsion* strength to prevent distortion and deflection during machining. The column provides the machining center with the y-axis linear movement.

Table. The table, which is mounted on the bed, provides the machining center with the z-axis linear movement.

Servo System. The servo system, which consists of servo drive motors, ball screws, and position feedback *encoders*, provides fast, accurate movement and positioning of the XYZ axis slides. The feedback encoders mounted on the ends of the ball screws form a closed-*loop* system which maintains consistent high-positioning *unidirectional* repeatability of +0.0001in (0.003mm).

Spindle. The spindle, which is programmable in 1r/min *increments*, has a speed range of from 20 to 6000 r/min. The spindle can be of a fixed position (horizontal) type, or can be a tilting/contouring spindle which provides for an additional axis.

Tool Changers. There are basically two types of tool changers, the vertical tool changer and the horizontal tool changer. The tool changer is capable of storing a number of preset tools which can be automatically called for use by the part program. Tool changers are usually *bidirectional*, which allows for the shortest travel distance to randomly access a tool[2]. The actual tool change time is usually only 3 to 5s.

MCU. The MCU allows the operator to perform a variety of operations such as programming, machining, *diagnostics*, tool and machine monitoring, etc. MCUs vary according to manufacturers' specifications; new MCUs are becoming more sophisticated, making machine tools more reliable and the entire machining operations less dependent on human skills.

Machine Axes. Machining centers have probably made the greatest impact in NC machining because of their ability to perform such a variety of machining operations on all sides of a workpiece with only one setup. The five-axis machining center indicates the axes that can be used when performing these machining operations and sequences.

x-axis	Linear movement
y-axis	Linear movement
z-axis	Linear movement
A-axis	Tilt/contour spindle
B-axis	Rotary table

Questions

1. How many types of machining centers are mentioned in the text?
2. List the main parts of computerized numerical control (CNC) for machining centers.
3. List the machining operations and sequences of five-axis machining center.

New Words and Expressions

1. pallet/ˈpælit/n. 托盘, 夹板, 棘爪, 货盘
2. shuttle/ˈʃʌtl/n. 穿梭, 往复运行工具; v. 使穿梭般来回移动
3. rotate/rəuˈteit/v. (使) 旋转, 转动
4. saddle/ˈsædl/n. 鞍状物, 床鞍, 滑鞍; v. 使负担
5. bedway/bedwei/n. 床身导轨; 滑板
6. quill/kwil/n. 衬套, 套管轴, 羽毛; vt. 卷在线轴上
7. torsion/ˈtɔːʃən/n. 扭转

8. encoder/inˈkəudə/n. 编码员, 编码器
9. loop/luːp/n. 圈, 环, 线圈, 环状物; v. 连成回路
10. unidirectional/juːnidiˈrekʃənl/a. 单向性的
11. increment/ˈinkrimənt/n. 增加, 增量, 增长
12. bidirectional/baidiˈrekʃənl/a. 双向的
13. diagnostic/ˌdaiəgˈnɔstik/a. 判断的, 诊断的

Notes

Glossary of Terms

1. servo system 伺服系统
2. cutter saddle 刀架
3. cylinder saddle 鞍形气缸座
4. a safety loop 保险圈
5. a wire loop 钢丝圈
6. loop a line 环路法连接线路
7. horizontal spindle 水平主轴
8. vertical spindle 垂直主轴
9. traveling-column 行程立柱
10. fixed-column 固定立柱
11. open-loop system 开环系统
12. close-loop system 闭环系统
13. feedback unit 反馈单元
14. machining center 加工中心
15. tool-storage 刀具存储
16. ball screw 滚珠丝杠
17. tool changer 换刀装置
18. machine control unit（MCU） 机床控制单元
19. revolving cutter 旋转刀具
20. flexible machining system 柔性制造系统

Reading Materials

The PC300 Mill

The HERCUS PC300 is a rugged machine and can perform milling, drilling, boring and tapping operations programmable in canned cycles on materials such as aluminium, brass, plastic, cast iron and steel.

Each of the three axes have wide dovetail slideways with adjustable gibs and anti-friction axis screws to give high stiffness and extended life. Both the y and x axes slides have stainless steel shields to protect them from swarf while the z axis slide is fully covered at all times.

The axis drives feature DC Servo motors with shaft encoders to give a reliable closed loop system. The DC spindle drive gives infinitely variable spindle speeds in two ranges which can be set by changing the toothed belt on the spindle head. The spindle has a No. 30 International Taper which enables a wide range of standard readily available tooling to be used programmable up to 10 tools.

Mechanical Specifications

Table Working Area	20in × 6. 5in(508mm × 165mm)
Longitudinal Travel—x axis	12in(304. 8mm)
Cross Travel—y axis	6. 75in(171. 45mm)
Vertical Travel—z axis	8. 75in (222. 25mm)
Spindle To Table —max	11. 75in (298. 45mm)
—min	3in(76. 2mm)
Maximum Rapid Traverse	150in/min (3810mm/min)
Spindle To Column-throat	6in (152. 4mm)
Table Tee Slots —number	3
—size	0. 44in(11. 18mm)
—spacing	1. 88in(47. 75mm)
Spindle Taper	No. 30 ISO (N. M. T. B)
Variable Spindle Speeds	0 ~ 3500r/min
—range 1	0 ~ 1100r/min
—range 2	0 ~ 3500r/min
Spindle Drive Motor DC	0. 5 HP (0. 37kW)
Electrical Input —single phase	110 V or 240 V,50 Hz or 60 Hz
Position Accuracy	± 0. 001in(± 0. 025mm)
Dimensions	L: 36in(914. 4mm)
	W: 34in(863. 6mm)
	H: 36in(914. 4mm)

ISO Standard M Codes

The number of M codes standardized by ISO is less compared to G codes in view of the direct control exercised by these on the machine tool. The ISO standard M codes are given in the following table(Table 26-1).

Table 26-1　ISO standard M codes

Code	Function	Code	Function
M00	Program stop, spindle and coolant off	M36 ~ M39	Unassigned
M01	Optional programmable stop	M40 ~ M45	Gear changes; otherwise unassigned
M02	End of program—often interchangeable with M30	M46 ~ M49	Unassigned
M03	Spindle on, CW	M50	Coolant supply No. 3 on
M04	Spindle on, CCW	M51	Coolant supply No. 4 on
M05	Spindle stop	M52 ~ M54	Unassigned
M06	Tool change	M55	Linear cutter offset No. 1 shift
M07	Coolant supply No. 1 on	M56	Linear cutter offset No. 2 shift
M08	Coolant supply No. 2 on	M57 ~ M59	Unassigned
M09	Coolant off	M60	Piece part change
M10	Clamp	M61	Linear piece part shift, location 1
M11	Unclamp	M62	Linear piece part shift, location 2
M12	Unassigned	M63 ~ M67	Unassigned
M13	Spindle on, CW + coolant on	M68	Clamp piece part
M14	Spindle on, CCW + coolant on	M69	Unclamp piece part
M15	Rapid traverse in + direction	M70	Unassigned
M16	Rapid traverse in – direction	M71	Angular piece part shift, location 1
M17 ~ M18	Unassigned	M72	Angular piece part shift, location 2
M19	Spindle stop at specified angular position	M73 ~ M77	Unassigned
M20 ~ M29	Unassigned	M78	Clamp non-activated machine bedways
M30	Program stop at end tape + tape rewind	M79	Unclamp non-activated machine bedways
M31	Interlock by-pass	M80 ~ M99	Unassigned
M32 ~ M35	Constant cutting velocity		

Unit Ten

Lesson 27　The Computer in Die Design

Text

The term CAD is alternately used to mean computer aided design and computer aided drafting. Actually it can mean either one or both of these concepts, and the tool designer will have occasion to use it in both forms.

CAD computer aided design means using the computer and *peripheral* devices to simplify and enhance the design process. CAD computer aided drafting means using the computer and peripheral devices to produce the *documentation* and *graphics* for the design process[1]. This documentation usually includes such things as preliminary drawings, working drawings, parts lists, and design calculations.

A CAD system, whether taken to mean computer aided design system or computer aided drafting system, consists of three basic components: (1) hardware, (2) software, and (3) users. The hardware components of a typical CAD system include a processor, a system display, a keyboard, a *digitizer*, and a *plotter*. The software component of a CAD system consists of the programs which allow it to perform design and drafting functions. The user is the tool designer who uses the hardware and software to simplify and enhance the design process.

The broad-based *emergence* of CAD on an industry-wide basis did not begin to materialize until the 1980's. However, CAD as a concept is not new. Although it has changed *drastically* over the years, CAD had its beginnings almost thirty years ago during the middle 1950's. Some of the first computers included graphics displays. Now a graphics display is an *integral* part of every CAD system.

Graphics displays represented the first real step toward bringing the worlds of tool design and the computer together. The plotters *depicted* in Figure, represented the next step. With the *advent* of the digitizing *tablet* in the early 1960's, CAD hardware as we know it today began to take shape. The development of computer graphics software followed soon after these hardware developments.

Early CAD systems were large, *cumbersome*, and expensive. So expensive, in fact, that only the largest companies could afford them. During the late 1950's and early 1960's, CAD was looked on as an interesting, but impractical novelty that had only limited *potential* in tool design applications. However, with the introduction of the silicon chip during the 1970's, computers began to take their place in the world of tool design.

Integrated circuits on silicon chips allowed full scale computers to be packaged in small *consoles* no larger than television sets[2]. These "mini-computers" had all of the characteristics of full scale computers, but they were smaller and considerably less expensive. Even smaller computers called microcomputers followed soon after.

The 1970's saw continued advances in CAD hardware and software technology. So much so that by the beginning of the 1980's, making and marketing CAD systems had become a growth industry. Also, CAD has been transformed from its status of impractical *novelty* to its new status as one of the most important inventions to date. By 1980, numerous CAD systems were available ranging in sizes from microcomputer systems to large minicomputer and mainframe systems.

Questions

1. What are the two different interpretations of the term CAD? What is the difference between these two concepts?

2. What are the three basic components included in a typical CAD system?

3. What are the hardware components included in a typical CAD system?

New Words and Expressions

1. peripheral/pəˈrifərəl/*a.* 边缘的, 外表面的, 周边的

2. documentation/ˈdɔkjumenˈteiʃən/*n.* 文件(编制), 记录, 提供文件

3. graphics/ˈɡræfiks/*n.* 制图法, 图解计算法

4. digitize/ˈdidʒitaiz/*vt.* 使(模拟值)数字化

5. plotter/ˈplɔtə/*n.* 绘图机

6. emergence/iˈməːdʒəns/*n.* 浮现, 显露, 出现, 冒出, 发生

7. drastic/ˈdræstik/*a.* 强有力的, 猛烈的; *n.* 烈性药物

8. integral/ˈintiɡrəl/*a.* 完整的, 完全的,

组成的; *n.* 总体, 整体

9. depict/diˈpikt/*vt.* 描写, 描述, 描绘

10. advent/ˈædvənt/*n.* 到来, 出现

11. tablet/ˈtæblit/*n.* 小片, 小块, 平板

12. cumbersome/ˈkʌmbəsəm/*a.* 繁重的, 笨重的, 麻烦的

13. potential/pəˈtenʃəl/*a.* 潜在的, 可能的; *n.* 势能, 位能, 潜力

14. console/ˈkɔnsəul/*n.* 仪表板, 控制台, 托架

15. novelty/ˈnɔvəlti/*n.* 新颖, 新奇

16. silicon chip 硅片

17. integrated circuit 集成电路

Notes

[1] CAD computer aided drafting means using the computer and peripheral devices to produce the documentation and graphics for the design process.

CAD 计算机辅助绘图是指用计算机和其外设来制作设计过程的文件和图样。

computer aided drafting 译为"计算机辅助绘图", peripheral devices 译为"外设装置(或外设)"。

[2] Integrated circuits on silicon chips allowed full scale computers to be packaged in small consoles no larger than television sets.

由于计算机所需的集成电路均制作在小小的硅片上，因此组装起来的全功能计算机与电视机一样大。

no larger than 译为"与……一样大"。

Glossary of Terms

1. disk operating system（DOS） 磁盘操作系统
2. microsoft disk operating system（MS-DOS）微软磁盘操作系统
3. operating system user's guide 操作系统用户指南
4. program and data files 程序和数据文件
5. internal and external command 内部和外部命令
6. system prompt 系统提示
7. format a diskette 磁盘格式化
8. diskcopy command 磁盘复制命令
9. erase（deletion）command 删除命令
10. create（change, remove）directory 建立（改变，移动）目录
11. hard disk drive（HDD） 硬盘驱动器
12. hard（soft）disk 硬（软）盘
13. floppy disk drive（FDD） 磁盘驱动器
14. standard keyboard 标准键盘
15. color display 彩色显示
16. printer operating procedures 打印机操作程序
17. cursor control 光标控制
18. application window 应用程序窗口
19. batch file 批处理文件
20. control（main, system）menu 控制（主，系统）菜单
21. drop-down menu 下拉式菜单
22. configuration system file 系统配置文件

Reading Materials

AutoCAD Menu

File Menu for AutoCAD

New...   Creates a new drawing file. See NEW.

Open...   Opens an existing drawing file. See OPEN.

Save   Saves the current drawing. See QSAVE.

Save As...   Saves an unnamed drawing with a file name or renames the current drawing. See SAVEAS.

Export...   Saves objects to other file formats. See EXPORT.

Printer Setup...   Displays the printer tab of the preferences dialog box. See PREFERENCES.

Print Preview   Shows how the drawing will look when it is printed or plotted. See PREVIEW.

Print. . .   Plots a drawing to a plotter, printer, or file. See PLOT.

Drawing Utilities   Drawing Utilities submenu

Send. . .   Sends current drawing file as an email attachment.

Exit   Exits AutoCAD. See QUIT.

Format Menu

Layer. . .   Manages layers. See LAYER.

Color. . .   Sets the color for new objects. See DDCOLOR.

Linetype. . .   Creates, loads, and sets linetypes. See LINETYPE.

Text Style. . .   Creates named styles. See STYLE.

Dimension Style. . .   Creates and modifies dimension styles. See DDIM.

Point Style. . .   Specifies the display mode and size of point objects. See DDPTYPE.

Multiline Style   Defines a style for multiple parallel lines. See MLSTYLE.

Units. . .   Controls coordinate and angle display formats and determines precision. See DDUNITS.

Thickness   Sets the current 3D thickness. See THICKNESS.

Drawing Limits   Sets and controls the drawing boundaries. See LIMITS.

Rename. . .   Changes the names of named objects. See DDRENAME.

Tools Menu

Spelling   Checks spelling in a drawing. See SPELL.

Display Order   Display order submenu

Inquiry   Inquiry submenu

Load Application. . .   Loads AutoLISP, ADS, and ARX applications. See APPLOAD.

Run Script. . .   Executes a sequence of commands from a script. See SCRIPT.

Display Image   Display image submenu

External Database   External database submenu

Object Snap Settings. . .   Sets running object snap modes and changes the target box size. See OSNAP.

Drawing Aids. . .   Sets drawing aids. See DDRMODES.

UCS   UCS submenu

Grips. . .   Turns on grips and sets their color. See DDGRIPS.

Selection. . .   Sets object selection modes. See DDSELECT.

Object Group. . .   Creates a named selection set of objects. See GROUP.

Tablet   Tablet submenu

Customize Menus. . .   Customizes menus. See MENULOAD.

Preferences. . .   Customizes the AutoCAD settings. See PREFERENCES.

Lesson 28　CAD/CAM

Text

Computer-aided design/computer-aided manufacturing (CAD/ CAM) refers to the integration of computers into the design and production process to improve productivity. The heart of the CAD/CAM system is the design terminal and related hardware, such as computer, printer, **plotter**, paper tape punch, a tape reader, and digitizer. The design is constantly monitored on the **terminal** until it is completed. A hard copy can be generated if necessary. A computer tape or other control medium containing the design data guides computer-controlled machine tools during the manufacturing, testing, and quality control.

The software for CAD/CAM is a collection of computer programs stored in the system to make the various hardware components perform specific tasks. Examples of software are programs developed to generate a NC tool path, to assemble a bill of materials, or to create nodes and elements on a finite element model. Some of these software packages are referred to as software modules and can be classified into four categories: (1) operating systems, (2) general-purpose programs, (3) application programs, and (4) user programs. Although there are other kinds of software, these are sufficient for an explanation of the complexities in developing a CAD/CAM system.

Operating systems are programs written for a specific computer or class of computers. For convenient and efficient operation, programs and data are available in the system's **memory**. The operating system is especially concerned with the input/output (I/O) devices like displays, printers, and tape punches. In most cases the operating system is supplied with the computer.

Although it may be argued that there are no general-purpose programs as such, some are more general than others. An example is a graphics program written in a high-level language like FORTRAN that allows the generation of geometric entities such as lines, circles, and **parabolas** and a combination of these to make designs. These designs may range from printed circuits to drill jigs and fixtures.

Application programs are developed for a special or specific purpose. The first language for specialized application was Automatically Programmed Tools (APT) in 1956. APT was developed to ease the job of NC programmers in developing input to NC machine tools, as illustrated in Figure 28-1. Other examples of application programs, relative to CAD/CAM, are programs developed specifically for the generation of finite element mesh and flat pattern development or "unbending" of sheet metal

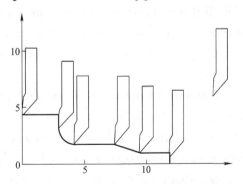

Figure 28-1　Example of CAD where lathe tool is called up to show a machining sequence

parts. These programs are usually purchased with the system or from a software supplier.

User programs in CAD/CAM are highly specialized packages for creating specific outputs. For example, a user program may automatically design a gear after the user inputs certain parameters like the number of teeth, pitch diameter, and so on. Another program may calculate optimum feeds and speeds, given cutter information, material, depth of cut, and so on. These programs are often developed by the user from a software module furnished by the supplier of general-purpose software. Not all CAD/CAM software packages have these user programs, even though considerable savings can be achieved with them.

Computer Graphics. The computer graphics system accumulates and stores physically related data identifying the precise location, dimensions descriptive text, and other properties of every design element. The design-related data help the user—operator perform complex engineering analysis, generate bills of materials, produce reports, and detect design *inconsistencies* before the part reaches manufacturing.

With computer graphics two-dimensional drawings can be made into three-dimensional wire frames and solid models.

Wire Frames. The simple wire frame plot is the least expensive form of geometrically displaying a model. It is useful to verify the basic properties of a shape and continuity of the model. However, when a complex model is developed, wire frame displays become inadequate. Solid models eliminate most of the problems of the wire frame.

Solid Modeling. There are three basic techniques for generating solid models: constructive solid geometry (CSG), boundary representation (B-Rep), and analytical solid modeling.

In the CSG approach, various geometric patterns such as *cylinders*, spheres, and *cones* are combined by Boolean algebra to create designs.

In the B-Rep method, a profile of the part is defined and then swept, either linearly or radially, and the enclosed area represents the solid form.

Analytical Method. This method is similar to the B-Rep but enhances the creation of finite element model during generation of the design. Commercial packages do not use strictly one method or another. As an example, CSG packages may use B-Rep techniques to generate initial patterns, while B-Rep or analytical packages may use Boolean algebra to subtract patterns, such as cylinders or cones, from a design to create a hole in the design.

Computer-aided manufacture (CAM) centers around four main areas: NC, process planning, robotics, and factory management.

Numerical Control. The importance of NC in the CAM area is that the computer can generate a NC program directly from a geometric model or part. At present, automatic capabilities are generally limited to highly symmetric geometries and other specialized parts. However, in the near future some companies will not use drawings at all, but will be passing part information directly from design to manufacturing via a data base. As the drawings disappear, so will many of the problems, since computer models developed from a common integrated data base will be used by both design and manufacturing[1]. This can be done even though the departments may be widely separated geographically, because in essence they will be no farther apart than the terminals on their respective desks.

Process Planning. Process planning involves the detailed planning of the production sequence from start to finish. What is relevant to CAM is a process planning system that is able to produce process plans directly from the geometric model data base with almost no human assistance.

Robots. Many advances are being made to integrate robots into the manufacturing system, as in on-line assembly, welding, and painting.

Factory Management. Factory management uses interactive factory data collection to get timely information from the factory floor. At the same time, it uses this data to calculate production priorities and dynamically determine what work needs to be done next to ensure that the master production schedule is being properly executed[2]. The system can also be directly modified to satisfy a specific need without calling in computer programming experts.

Questions

1. What is the heart of the CAD/CAM system?
2. How many categories can software packages for CAD/CAM be divided into and what are these?
3. What are the differences between wire frames and solid modeling in computer graphics?
4. List four main areas of computer-aid manufacture.
5. Why some companies will not use drawing at all in the near future?

New Words and Expressions

1. plotter/ˈplɔtə/*n.* 绘图机，标图员，标绘器
2. terminal/ˈtəːminl/*a.* 终端的，极限的，期终的；*n.* 末端，末尾
3. memory/ˈmeməri/*n.* 内存
4. parabola/pəˈræbələ/*n.* 抛物线
5. inconsistency/inkənˈsistənsi/*n.* 不一致，不合理
6. cylinder/ˈsilində/*n.* 圆柱（体）
7. cone/kəun/*n.* 锥体，锥形；*v.* （使）成

锥形
8. a bill of material 材料表（清单）
9. be referred to as 被称为……
10. be concerned with 与……有关
11. printed circuits 印制电路
12. pitch diameter 节圆直径
13. software package 软件包
14. Boolean algebra 布尔代数
15. in essence 本质上

Notes

[1] As the drawings disappear, so will many of the problems, since computer models developed from a common integrated data base will be used by both design and manufacturing.

由于不再使用图样，从而使因使用图样所带来的许多问题也会随之消失，这是因为此时设计和制造所使用的计算机模型都是由一个通用的集成数据库形成的。

so will many of the problems 是一个倒装句，这种以 so（此外还有 neither 和 nor）开头的句子中还常用省略式句型，例如：Two electrons will be repelled from each other and so will two nuclei. 两个电子相互排斥，两个原子核也是如此。

[2] At the same time, it uses this data to calculate production priorities and dynamically determine what work needs to be done next to ensure that the master production schedule is being properly executed.

同时，利用这些数据计算出制造的先后次序、动态地确定下一步要做的工作，从而确保正常地执行标准制造程序。

句中 to calculate production priorities and dynamically determine what work needs to be done next 为宾语补足语；to ensure that the master production schedule is being properly executed 为目的状语从句。

Glossary of Terms

1. FMS（flexible manufacturing system）柔性制造系统
2. CNC（computer numerical control）计算机数字控制
3. PTP（point to point）点到点的定位方式
4. spindle with sensor 传感轴
5. revised feed signal 反进给信号
6. default selection mode 默认选择模式
7. MCU（machine control unit）加工控制单元
8. closed-loop system 闭环控制系统
9. ACS（adaptive control system）自动补偿系统
10. CRT（cathode-ray tube）显像管
11. geometric modeling 建立几何模型
12. process planning 制订工艺过程
13. CIM（computer integrated manufacturing）计算机集成制造
14. vertical stroke 垂直行程
15. limit switch 限位开关
16. servo control 伺服控制（系统）
17. font file 字体文件
18. system variable 系统变量
19. plot preview 出图预览

Reading Materials

Robots

The word robot was first introduced by a Czech playwright, Karel Capek, in 1917. Robot, the Czech word meaning servitude and drudgery, is still the concept that is prevalent for robots today. Although the word is now commonplace, it was not until 1973 that the first commercially available, minicomputer-controlled robot was built for Cincinnati Milacron, a machine tool builder.

The Robotic Industries Association (RIA) recently offered this definition of an industrial robot: "An industrial robot is a reprogramable, multifunctional manipulator designed to move material, parts, tools, or specialized devices through variable programmed motions to accomplish a variety of tasks." As the definition implies, the robot must have the ability to adapt to many different kinds of jobs in many different industries and perform these jobs with some degree of dexterity and flexibility in motions. In contrast, fixed or special-purpose automation is less universal and includes machines designed to do one particular job in one specific industry. Although industrial robots may vary widely, they are all made up of three basic components: a manipulator, power supply, and control system. These components may be assembled in one integral unit or separated into individual components connected by pneumatic, hydraulic, or electrical "umbilical cords".

The manipulator is an assembly of axes capable of motion in various directions. The "wrist" located at the end of the robot arm has from one to three degrees of freedom, depending on the model or make. There is no standard terminology in the industry as yet, but these degrees of freedom may be termed pitch, yaw, and roll axes or bend, yaw, and swivel axes.

The manipulator is powered by hydraulic or pneumatic cylinders or rotary actuators, and electric or hydraulic motors are used to power the various axes of motion. Also attached to the manipulator are feedback devices that sense and measure the position and, in some cases, the velocity of each of the various axes of motion and send this information to the control systems for use in coordinating the robot motions.

The power supply is the source of energy used to move and regulate the robot's drive mechanisms. The energy comes from three sources: electric, hydraulic, and pneumatic. Electric drives are clean and quiet with a high degree of accuracy and repeatability. They also offer a wide range of payload capacity, accompanied by an equally wide range of costs. Hydraulic drives, today's most popular, have high payload capacities, and are relatively easy to maintain. They are, however, rather expensive and not as accurate as either the electric or pneumatic drives. Pneumatic drives, although limited to smaller payloads, are relatively inexpensive, fast, and reliable.

Robots may be classified by work envelopes. The work envelope may be defined as all the points in space that can be touched by the robot's arm or tool mounting plate in three-dimensional space. The work envelope is dependent on the three major axes of motion: a vertical lift stroke, an in-and-out reach stroke of the arm, and a rotational or traverse motion about the vertical lift axes of the robot.

The robot control unit (RCU) is termed the brains of the robot. It sequences and coordinates the motion of the various axes of the robot and provides interlocking communication with external devices and machines. Controls may range from simple stepping-drum types or sequences to microcomputer-based types for the sophisticated robots complete with various memory devices.

System Fundamentals and Information to Pro/E

Once Pro/E has successfully been started, the workstation screen is divided into the following areas:

Main working window—where most graphics work is done.

Pop up menus—where Pro/E modes and commands are chosen.

Message window—where system operation information and prompts are shown.

Menus and Dialog Boxes provide an option driven interface mechanism for creating and redefining features in Pro/E. You can create or redefine a feature by providing Pro/E with certain information blocks called "options" and "elements" using the following tools:

FEATURE OPTION MENUS—Select specific feature "options", such as extrude, revolve, sweep, etc., and whether the feature will be solid or thin. Options can not be redefined.

FEATURE ELEMENT MENUS—Controls the process of defining the elements such as Direction and Depth, required to create the feature. Elements can be redefined.

DIALOG BOXES—Provide a uniform method of displaying and changing the feature elements and their current status, obtaining information regarding the feature and references created, as well as "previewing" and finalizing creation of the feature geometry.

When creating a feature, you must specify values for its options and elements. The options specified will determine the elements required to complete the feature.

System messages appear in the message window, at the bottom of the main work window. The messages have the following functions:

Provide a on-line help message about a menu being chosen. These appear in yellow at the bottom of the message window.

1. Provide information about the status of an operation. These appear in white letters as they scroll down the message window.

2. Query for a additional information to complete a command.

If Pro/E is providing information, proceed with another function and the message will disappear.

If Pro/E is asking for additional information, provide the information by entering it via the keyboard or indicating it with the mouse. When input is needed by Pro/E, a bell will sound to notify you. When data is required to be entered, a prompt will appear in the message window ending with a colon. When Pro/E requires data input all other functions are temporarily disabled until entering is complete.

The MAIN menu contains access to the major modes of Pro/E operation and other system related functions. These menu options are:

Mode—This menu item allows you to access major Pro/E modes. It brings up the MODE menu that allows you to choose the mode you are going to work.

Environment—This menu item affects the operation modes. It is used to change display mode settings, prompt bell status, and file storage settings.

Info—This menu item lets the user access information functions without having to quit your current Pro/E action.

Misc—This menu item lets the user access the workstation's operating system functions without leaving Pro/E, and re-run journal and training files in the Pro/E environment.

Exit—This menu item ends the Pro/E working session.

Quit Window—This menu item lets the user quit from the current window being used.

Change Window—This menu item lets the user change the window being worked in, from one to another.

View—This menu item is used to change the view of an object.

Menu items are chosen using the mouse. To choose an item, move the cursor over the menu item. That item will appear to protrude from the menu list. Click the left mouse button to activate a selection. The menu item will now either highlight in black or appear to protrude from the menu, and the function will be executed.

As you scroll the cursor up and down the menu items, a one-line help message describing the function that is highlighted appears in the system message window.

In all Pro/E modes (Sketcher, Part, Assembly, Drawing, etc.), files can be created, retrieved, imported, and listed through an ENTERSECTION, ENTERPART, ENTERASSY, or ENTERDRAWING menu, depending on what mode you are in. In order to work in these modes, a new file on which to work must be created or an existing file retrieved.

Pro/E lets you work in a multiple window environment. In this way, you can easily change your working window from one to another. You can also refer to the information in a window without activating it.

Within Pro/E, operating environment settings can be changed by the user. The default settings can be changed by entering new default options in a file named CONFIG. PRO in the current working directory. In this way, each user can customize their own environment.

The Model Tree can be used to display all the features in your model configure information columns for each feature. Information columns can be added and removed and the format changed to personal desires. The Model Tree can be disabled by toggling off the check mark in the ENVIRONMENT menu.

Lesson 29　Rapid Prototyping and Manufacturing

Text

Prototyping (*prototype*) is "the *original* thing in relation to any copy, *imitation*, *representation*, later specimen or improved form" (taken from Webster's dictionary). Rapid prototyping is "fabrication of a physical, three-dimensional part of *arbitrary* shape directly from a numerical *description* (typically a CAD *model*) by a quick, highly automated and totally flexible processes", "Rapid Prototyping and Report", October 1992.

In recent years and constantly being updated at the time of this writing, several types of rapid prototyping and manufacturing (RP&M) have *emerged*. The technologies developed include stereolithography (SL), selective laser sintering (SLS), fused deposition modeling (FDM), laminated object manufacturing (LOM), and three dimensional printing (3-D printing). They are all capable of generating physical objects from computer aided design (CAD) databases.

With these technologies, product development times are substantially decreased and flexibility for manufacturing a variety types and sizes of products is improved.

The Feature of Rapid Prototyping and Manufacturing. RP&M process consists of two steps: step 1, a part is first modeled by a 3-D solid *geometric* modeller, and then is sliced into a series of parallel 2-D cross-section layers in computer; step 2, the *datum* of the 2-D layers is directly used to instruct the machine for producing the part layer by layer from bottom to top. A common important feature of RP&M is that the prototype part is produced by assigning materials rather than removing materials[1].

The benefits from applying the technology to improve product development are in the following three aspects:

(1) Design engineering

Designers use CAD to generate *visual* models of actual complex products, that is prototypes of products, in short time, therefore engineers can evaluate a design very quickly.

An RP&M prototype can be produced quickly without substantial tooling and labor cost, and the product quality can be improved within the limited time frame and with *affordable* cost[2].

(2) Manufacturing

By providing a physical product at design stage we can speed up process planning and tooling design, reduce problems in interpreting the blue prints on the shop floor.

(3) Marketing

By *demonstrating* the concept, design ideas, as well as the company's ability to produce it as a prototype, we can gain customer's feedback for design *modifications* in timely manner, and promote product sales.

Stereolithography (SL), which is rapid prototyping by laser curing a photocurable liquid, was launched commercially by 3D systems Inc. in 1987.

SL *apparatus* creates the prototype by tracing layer cross-sections on the surface of the liquid resin photopolymer pool with a laser beam. As show in Figrue 29-1, the laser moves as a point source across the surface of the liquid, first curing the bottom *slice* of the object. This slice moves down on an elevator by 50 to 375 microns (0.002 to 0.015in, 1in = 25.4mm), depending on desired accuracy. The next layer is then photocured, also fusing to the one below.

The photocurable liquid resin was developed for printing and furniture *lacquer/ sealant*. The laser provides direct energy. With the energy, the original *vinyl monomers* (small molecules) are polymerized into large molecules, which get much more strength from this process.

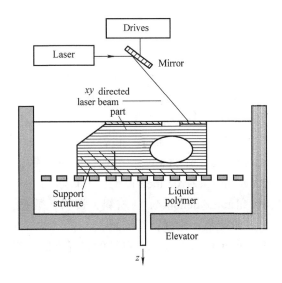

Figure 29-1　Stereolithography (SL), based on commercially published brochures of 3D systems Inc

Unlike the contouring or *zigzag* cutter movement used in CNC machining, in order to create any individual layer, the laser traces out the boundaries of layer first. This is called bordering, imagine a large elastic *band* or *loop* lying on the surface. Secondly, *hatched* areas are filled in, causing the final *gelling* and solidification. After each layer is formed, the scanning moves to the next layer to draw a new layer on the top of the precious one. In this way, the models built layer by layer from bottom to top. However, some careful process planning is needed to create the accuracy of only a few thousandths of an inch (25.4mm).

When all layers are completed, the prototype is about 95% cured. Post-curing is needed to completely solidify the prototype. This is done in a *fluorescent* oven where *ultraviolet* floods the object (prototype).

Questions

　　1. What is prototyping?
　　2. What are the features of rapid prototyping and manufacturing?
　　3. What is stereolithography (SL)?
　　4. Why do we need rapid prototyping and manufacturing?

New words and Expressions

1. prototype/ˈprəutətaip/*n*. 原型，初型，模型；试制型式，样品

2. original/əˈridʒənəl/*a*. 最初的，原先的，新颖的；*n*. 原作品，原物

3. imitation/imiˈteiʃən/*n*. 模仿，仿造

4. representation/reprizenˈteiʃən/*n*. 代表，

表示，描写

5. arbitrary/ˈɑːbitrəri/*a*. 随意的，任意的，专断的

6. description/disˈkripʃən/*n*. 描写，叙述；图说，说明书；作图，绘制

7. model/ˈmɔdl/*n*. 模型，原型，样品，设计图；*vt*. 仿造，设计

8. emerge/iˈməːdʒ/*vi*. 浮现，出现，发生

9. geometric/dʒiəˈmetrik/*a*. 几何学的

10. datum/ˈdeitəm/*n*. 材料，资料；数据，给定数；论据；基准面

11. visual/ˈviʒuəl/*a*. 可见的，视觉的，光学的

12. afford/əˈfɔːd/*vt*. （常接在 can, could, be able to 后）买得起，抽得出（时间）；提供

13. demonstrate/ˈdemənstreit/*vt*. 论证，证明，说明，示范

14. modification/ˌmɔdifiˈkeiʃən/*n*. 修改，改变，改进；限制，变形；修饰

15. apparatus/ˌæpəˈreitəs/*n*. 器械，仪器，设备，装置；机构，机关

16. slice/slais/*n*. 切片，分层；*vt*. 把……切成薄片

17. lacquer/ˈlækə/*n*. 漆；*vt*. 上漆

18. sealant/ˈsiːlənt/*n*. 密封剂

19. vinyl/ˈvainil/*n*. 乙烯基；~ resin 乙烯基树脂

20. monomer/ˈmɔnəmə/*n*. 单体

21. zigzag/ˈzigzæg/*n*. Z 字形，锯齿形，之字形；*a*. 曲折的

22. band/bænd/*n*. 带，波带，能带

23. loop/luːp/*n*. 圈，环，线圈，环状物；*vt*. 使成圈，做环状运动

24. hatch/hætʃ/*n*. 孔，天窗，图画阴影线；*vt*. 画阴影线

25. gel/dʒel/*n*. 凝胶；*vi*. 胶化

26. fluorescent/fluəˈresnt/*a*. 荧光的，发荧光的

27. ultraviolet/ˌʌltrəˈvaiəlit/*a*. 紫外（线）的，产生紫外线的；*n*. 紫外线辐射

28. in（with）relation to 关于，涉及，与……有关

29. be filled in 填充，填满，把……插进去

30. be capable of 能够，可以

31. trace out 描出……轨迹，轨迹为

32. in this way 用这种方法

33. post-curing 辅助固化，后固化，二次固化

34. photocurable 能光固化的

35. layer by layer 一层一层地

36. from bottom to top 从底到顶

37. prototype workpiece 样件

Notes

[1] A common important feature of RP&M is that the prototype part is produced by assigning materials rather than removing materials.

快速原型制造的一个共有的重要特征是通过分配材料而不是去除材料来生产原型零件。

句中 that the prototype part is produced by assigning materials rather than removing materials 为表语从句，可译为 "通过分配材料而不是去除材料来生产原型零件"，rather than 作 "而不是" 解。

[2] An RP&M prototype can be produced quickly without substantial tooling and labor cost, and the product quality can be improved within the limited time frame and with affordable cost.

没有实质上的工具和劳动成本，快速原型制造能快速制造出原型，并且还能在限定的时间范围内用可能提供的成本改进产品的质量。

句中 without substantial tooling and labor cost 和 within the limited time frame and with affordable cost 为介词短语，分别在句中作状语。without 作"没有"解，within 作"在……内"解，而 with 作"用"解。

Glossary of Terms

1. rapid prototyping（RP） 快速原型（成形）

2. rapid prototyping and manufacturing（RP&M） 快速原型（成形）制造

3. three dimensional printing（3-D printing） 三维打印

4. freeform fabrication 自由制造

5. desk-top manufacturing 桌面制造

6. stereolithography（SL） 光固化立体造型

7. stereolithography apparatus（SLA） 光固化成形机

8. selective laser sintering（SLS） 选域（选区）激光烧结

9. fused deposition modeling（FDM） 熔融沉积成形

10. laminated object manufacturing（LOM） 分层实体制造（也叫叠层制造）

11. laser technology 激光技术

12. dislodge forming 去除成形（车、铣、刨、磨、钻等）

13. additive forming 添加成形（如快速成形）

14. forced forming 受迫成形（如铸造、锻造、粉末冶金等）

15. growth forming 生长成形（利用生物材料的活性，如"克隆"）

16. computed tomography（CT） 计算机 X 射线分层造影技术，断层扫描

17. nuclear magnetic resonance（NMR） 核磁共振

18. high temperature sintering 高温烧结

19. direct metal deposition（DMD） 直接金属沉积

20. rapid tooling 快速模具（工具）制造

21. product quality 产品质量

Reading Materials

Selective Laser Sintering（SLS）

Selective laser sintering（SLS）was first commercialized by DTM corporation, and is rapid prototyping by laser tracing the shape of the part to be modeled in a thin layer of polymer or ceramic powder, and sintering（softening and bonding）of the powder.

In this process the laser moves as a point source across the surface of the powder, first sintering the bottom slice of the desired object. A roller spreads more power and a second layer is sintered, also fusing to the one below. This process is repeated over layers of powder. Figure 29-2 shows the working principle of SLS.

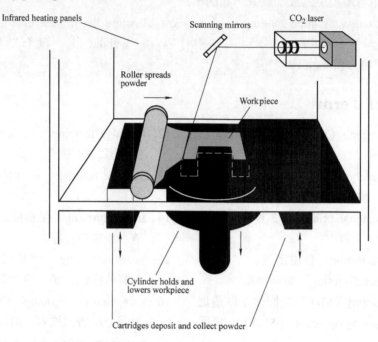

Figure 29-2   Selective laser sintering (SLS), based on
commercially published brochures from the DTM corporation

In many respects, SLS is similar to SL except that the laser is used to sinter and fuse powder rather than photocure a polymeric liquid, and a thin layer of fusible powder is heated by infrared heating panels. Otherwise comparing with SL, this process can rely on the supporting strength of the unfused powder around the partially fused object. Therefore, support columns for any overhanging parts of the component are not needed. This allows the creation of rather delicate, lacelike objects. Nevertheless SLS parts have a rough, grainy appearance from the sintering process, and it is often preferable to hand smooth the surfaces.

Laminated Object Manufacturing (LOM)

Laminated object manufacturing (LOM) was developed by Helisys Inc., and was first offered commercially in the period from 1987 to 1990, and is rapid prototyping by laser cutting the top layer of a stack of paper, each layer of which is glued down.

LOM processes produce parts from bonded paper, plastic, metal, or composite sheet stock. LOM machines bond a layer of sheet materials to a stack of previously formed laminations, and then a laser beam cuts the outline of the part cross-section generated by CAD to the required shape. The layers can be glued or welded together and the excess materials of every sheet either is removed by vacuum suction or remains as the next layer's support. This process is repeated by using very thin layers of materials. Figure 29-3 shows the working principle of LOM.

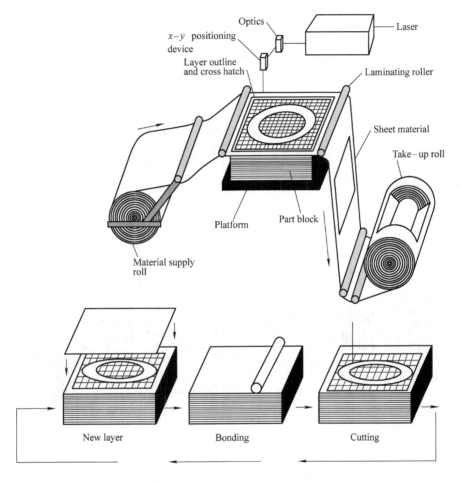

Figure 29-3　Laminated object manufacturing (LOM), based on
commercially published brochures of Helisys Inc

For larger components especially in the automobile industry, LOM is often preferred to the SL or SLS processes.

Fused Deposition Modeling (FDM)

Fused Deposition Modeling (FDM), which constructs parts based on deposition of extruded thermoplastic materials, was developed by Stratasys Inc. , and is executed on machines called the FDM 1620, 2000, or 8000 series.

The Fused Deposition Modeling process (FDM) is that the thermoplastic modeling material is fed into the temperature-controlled FDM extrusion head and heated to a liquid state. The head extrudes and deposits the material in ultrathin layers onto a fixtureless base.

Figure 29- 4 shows that the material is supplied as a filament from a spool. The overall geometry and system are reminiscent of icing a cake. The ribbon through an exit nozzle is guided by computer, and the viscous ribbon of polymer is gradually built up from a fixtureless base plate. In terms of motion control, FDM is more similar to CNC machining than SL or SLS. For simple parts, there is no need for fixing, and materials can be built up layer by layer. The creation of more complex parts with inner cavities, unusually sculptured surfaces, and overhanging features does require a support base, but the supporting materials can be broken away by hand, thus requiring minimal finishing work.

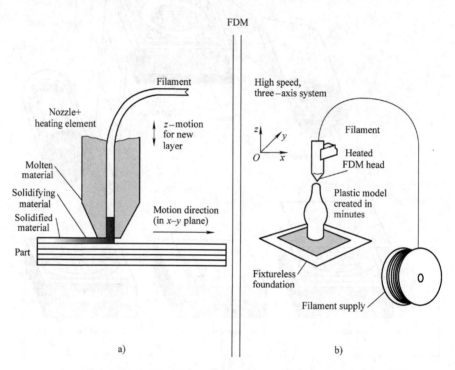

Figure 29-4　Fused deposition modeling (FDM), based on
published brochures of Stratasys Inc

a) Principle　b) System

Lesson 30　Advanced Manufacturing Technology

Text

Flexible Manufacturing. In the modern manufacturing *setting*, *flexibility* is an important characteristic. It means that a manufacturing system is *versatile* and *adaptable*, while also capable of handling relatively high production runs. A flexible manufacturing system is versatile in that it can produce a variety of parts. It is adaptable because it can be quickly *modified* to produce a completely different line of parts.

A flexible manufacturing system (FMS) is an *individual* machine or group of machines served by an automated materials *handling* system that is computer controlled and has a tool handling capability[1]. Because of its tool handling capability and computer control, such a system can be continually *reconfigured* to manufacture a wide variety of parts. This is why it is called a flexible manufacturing system. An FMS typically *encompasses*:

(1) Process equipment, e. g. , machine tools, assembly stations, and *robots*

(2) Material handling equipment, e. g. , robots, *conveyors*, and AGVS (automated guided vehicle system)

(3) A communication system

(4) A computer control system

Flexible manufacturing represents a major step toward the goal of fully integrated manufacturing. It involves integration of automated production processes. In flexible manufacturing, the automated manufacturing machine (i. e. , lathe, mill, drill) and the automated materials handling system *share instantaneous* communication *via* a computer network[2].

Flexible manufacturing takes a major step toward the goal of fully integrated manufacturing by integrating several automated manufacturing concepts:

(1) Computer numerical control (CNC) of individual machine tools

(2) *Distributed* numerical control (DNC) of manufacturing system

(3) Automated material handling systems

(4) Group technology (families of parts)

When these automated processes, machines, and concepts are brought together in one integrated system, an FMS is the result. Humans and computers play major roles in an FMS. The amount of human labor is much less than with a manually operated manufacturing system, of course. However, humans still play a *vital* role in the operation of an FMS. Human tasks include the following:

(1) Equipment troubleshooting, *maintenance*, and repair

(2) Tool changing and setup

(3) Loading and unloading the system

(4) Data input

(5) Changing of parts programs

(6) Development of programs

Flexible manufacturing system *components*. An FMS has four major components: Machines tools; Control system; Materials handling system; Human operators.

Questions

1. What is an FMS?
2. What are the human tasks in an FMS?
3. What are the four major components of an FMS?
4. Why do we need advanced manufacturing technology?

New Words and Expressions

1. setting/'setiŋ/n. 安装，装置，定位
2. flexibility/fleksə'biliti/n. 柔性，适应性，灵活性
3. versatile/'vəːsətail/a. 多方面的，多用途的，万用的，通用的
4. adaptable/ə'dæptəbl/a. 适合的，可适应的
5. modified/'mɔdifaid/a. 改良的，改进的；变形的，变性的
6. individual/indi'vidjuəl/a. 单独的，个别的；n. 个体
7. handling/'hændliŋ/n. 掌握，操作，处理
8. reconfigure/riːkən'figə/v. 重新装配，改装
9. encompass/in'kʌmpəs/vt. 包围，包含，完成
10. robot/'rəubɔt/n. 自动机，机器人
11. conveyor/kən'veiə/n. 运送者，传送装置，传送带
12. share/ʃɛə/n. 共享，参与；v. 分享，共有
13. instantaneous/instən'teiniəs/a. 同时的，立即的，瞬间的
14. via/'vaiə/prep. 经由，经过
15. distribute/dis'tribju(ː)t/vt. 分发，分配，分布，分类；v. 分发
16. vital/'vaitl/a. 生命的，主要的，有生命力的；n. 生机，要害
17. maintenance/'meintənəns/n. 维持，保持，支持，维修
18. component/kəm'pəunənt/n. 成分；a. 组成的，构成的
19. troubleshooting 故障检修，排除故障
20. be vital to 是……所必需的
21. a wide variety of 各种各样的，各种类型的
22. play a vital role in (play major roles in) 在……中起重要（主要）作用
23. a major step toward 迈向……的重要一步
24. production runs 流水线生产
25. a communication system 通信系统
26. a computer control system 计算机控制系统

Notes

[1] A flexible manufacturing system (FMS) is an individual machine or group of machines served by an automated materials handling system that is computer controlled and has a tool handling capability.

柔性制造系统（FMS）是由计算机控制并具有工具运送能力的自动化物料输送系统为之服务的一台机器或一组机器。

句中 that is computer controlled and has a tool handling capability 为由 that 引导的定语从句，修饰 an automated materials handling system。翻译时先翻译从句，后翻译主句。

[2] In flexible manufacturing, the automated manufacturing machine (i. e., lathe, mill, drill) and the automated materials handling system share instantaneous communication via a computer network.

在柔性制造过程中，自动化的制造机器（即车床、铣床、钻床）和自动化物料输送系统同时共享经由计算机网络传送的信息。

句中 in flexible manufacturing 为介词短语作状语。the automated manufacturing machine (i. e., lathe, mill, drill) and the automated materials handling system 为本句的主语。

Glossary of Terms

1. man-machine engineering（MME） 人机工程

2. International Standard Organization（ISO） 国际标准化组织

3. group technology（GT） 成组技术

4. advanced manufacturing technology 先进制造技术

5. flexible manufacturing cell（FMC） 柔性制造单元

6. flexible manufacturing system（FMS） 柔性制造系统

7. automated guided vehicle system（AGVS） 自动引导小车系统

8. computer-integrated manufacturing（CIM） 计算机集成制造

9. agile manufacturing（AM） 敏捷制造（灵捷制造）

10. virtual manufacturing（VM） 虚拟制造

11. environmentally conscious design and manufacturing（ECDM） 环保设计与制造

12. environmental engineering 环境工程

13. nanomaterial 纳米材料

14. nanotechnology 纳米技术

15. reverse engineering（RE） 反求工程

16. computer-aided process planning（CAPP） 计算机辅助工艺规程设计

17. computer-aided quality control（CAQC） 计算机辅助质量控制

18. computer-integrated production management（CIPM） 计算机集成生产管理

19. concurrent engineering（CE） 并行工程

20. just-in-time（JIT） 即时生产（精益生产、准时生产）

21. clean production（CP） 清洁化生产

22. lifecycle engineering（LCE） 生命周期工程

23. design for environment（DFE） 面向环境的设计

24. green product design（GPD） 绿色产品设计

25. intelligent manufacturing system（IMS） 智能制造系统

26. multi-media technology（MMT） 多媒体技术

27. chemical vapor deposition（CVD） 化学气相沉积

28. computer numerical control（CNC） 计算机数控

29. direct numerical control（DNC） 直接数控

Reading Materials

Computer Integrated Manufacturing (CIM)

Computer integrated manufacturing (CIM) is the term used to describe the modern approach to manufacturing. Although CIM encompasses many of the other advanced manufacturing technologies such as computer numerical control (CNC), computer-aided design/computer-aided manufacturing (CAD/CAM), robotics, and just-in-time (JIT) delivery, it is more than a new technology or a new concept. Computer integrated manufacturing is an entirely new approach to manufacturing, a new way of doing business.

To understand CIM, it is necessary to begin with a comparison of modern and traditional manufacturing. Modern manufacturing encompasses all of the activities and processes necessary to convert raw materials into finished products, deliver them to the market, and support them in the field. These activities include the following:

(1) Identifying a need for a product.

(2) Designing a product to meet the needs.

(3) Obtaining the raw materials needed to produce the product.

(4) Applying appropriate processes to transform the raw materials into finished products.

(5) Transporting products to the market.

(6) Maintaining the product to ensure proper performance in the field.

Fully integrated manufacturing firms realize a number of benefits from CIM:

(1) Product quality increases.

(2) Lead times are reduced.

(3) Direct labor costs are reduced.

(4) Product development times are reduced.

(5) Inventories are reduced.

(6) Overall productivity increases.

(7) Design quality increases.

Agile Manufacturing (AM)

Agile manufacturing is a means of thriving in an environment of continuous change, by managing complex inter and intra-firm relationships through innovations in technology, information and communication, organizational redesign and new marketing strategies.

There are four principles of agility:

(1) To organize to master change. "An agile company is organized in a way that allows it to thrive on change and uncertainty".

(2) To leverage the impact of people and information. In an agile company, knowledge is valued, innovation is rewarded, and authority is distributed to the appropriate level of the organization. There is a climate of mutual responsibility for joint success.

(3) To cooperate to enhance competitiveness. Cooperation internally and with other companies is an agile competitor's operational strategy of first choice.

(4) To enrich the customer. An agile company is perceived by its customers as enriching them in a significant way, not only itself.

From everything mentioned above, we can conclude that agile manufacturing involves more than just manufacturing. It involves the firm's organizational structure, the way in which the firm treats its people, the partnerships with other organizations, and the relationship with customers.

Nanomaterial and Nanotechnology

Nanomaterials and nanotechnology have become a magic word in modern society. Nanomaterials represent today's cutting edge in the development of novel advanced materials which promise tailor-made functionality and unheard applications in all key technologies. So Nanomaterials are considered as a great potential in the 21th century because of their special properties in many fields such as optics, electronics, magnetics, mechanics, and chemistry. These unique properties are attractive for various high performance applications. Examples include wear resistant surfaces, low temperature sinterable high-strength ceramics, and magnetic nanocomposites. Nanostructured materials present great promises and opportunities for a new generation of materials with improved and marvelous properties.

What is nanotechnology? It is a term that entered into the general vocabulary only in the late 1970's, mainly to describe the metrology associated with the development of X-ray, optical and other very precise components. We defined nanotechnology as the technology where dimensions and tolerances in the range $0.1 \sim 100nm$ (from the size of the atom to the wavelength of light) play a critical role.

Appendix

Appendix A　Translated Texts（课文参考译文）

第1单元

第1课　工具材料

在选择特定的工具材料时，应依据工具在工作时所需要的力学性能而定。只有在对工具的功能、需求进行细致的研究与评估后，才能确定应选材料。在实际应用中，许多情况下有多种材料能够满足工具的这种需求，然而，往往影响最终决定的却是材料的利用率和经济性。

主要的工具材料可分为三大类：钢铁材料、非铁金属材料与非金属材料。钢铁工具材料是以铁作为基体，包括工具钢、合金钢、碳钢和铸铁。非铁金属材料则不以铁作为基体，它包括铝、镁、锌、铅、铋、铜及各种合金。非金属材料不含金属基体，包括木头、塑料、橡胶、环氧树脂、陶瓷和金刚石。为了正确选择工具材料，你应当掌握材料的一些物理性能和力学性能，以便确定所选材料对工具的功能和操作会有何影响。

物理性能和力学性能是控制材料在特定状况下如何变化的特性。物理性能是材料的本身属性，如果不改变材料本身，这些性能不会发生永久变化。物理性能包括：质量、颜色、导热性与导电性、热膨胀率和熔点。力学性能是指材料经热加工或冷加工发生永久变化的性能。力学性能包括：强度、硬度、耐磨性、韧性、脆性、塑性、柔软性、延展性和弹性模量。

从应用的角度来看，像金属、塑料和木材这些常用材料在加工和成形为所需形状时会用到工具钢。从化学组成来说，工具钢是能够被淬硬并能经过回火变韧的碳素合金钢。工具钢应具有的性能还包括较高的耐磨性和硬度、良好的耐热性以及加工材料时所需的足够的强度。在某些情况下，尺寸稳定性也非常重要。工具钢在使用上还应满足经济性和加工工艺性要求，以便加工成所需要的工具形状。

由于存在一些特殊性能的要求，工具钢在电炉内冶炼时需要有严格的冶金质量控制，要最大限度地降低钢中的气孔、偏析、杂质及非金属夹杂物等的含量，并要通过仔细的宏观和微观检测，从而确保所炼制的工具钢达到严格的技术要求。

尽管工具钢在所有钢产品中仅占相对较小的比例，但是由于工具钢用于其他钢产品和工程材料的生产，因而具有战略性地位。工具钢的一些应用还包括钻头、深拉深模具、剪切刃、冲头、挤压模具和切削刀具。

对于某些应用场合，尤其在高速切削领域，使用烧结碳化物这样的材料相对工具钢来说更经济。烧结碳化物材料特有的性能源自于它的高硬度和高抗压强度。在工业领域中，其他工具钢材料的使用也越来越普遍。

第2课　工具钢的热处理

热处理的目的是通过加热金属或合金到一定的温度，然后以各种速度冷却去改变其组织结构来控制金属或合金的性能。这种加热和冷却控制相结合的方法不仅决定了材料中微观组织的分布和性质（进而决定了该材料的性能），而且也决定了材料内部晶粒的大小。

通过热处理可以使金属或合金获得所希望的性能，热处理的各种目的如下：

1．消除冷加工后的变形。

2．去除拉深、弯曲或焊接产品的内应力。

3．提高材料的硬度。

4．改善材料的可切削性能。

5．改善刀具的切削性能。

6．提高耐磨性。

7．软化金属材料，如退火工艺。

8．提高或改变材料的性能。如耐腐蚀性能，耐热性能，磁性或其他所需的性能。

钢铁材料的热处理。 在工具钢中，铁是主要的化学成分，加入碳的目的是为了提高钢的硬度。为了改变普通碳钢的性能，例如在油或空气中硬化的能力，提高耐磨性、韧性以及在淬火时的安全性，需要在钢中加入各种合金元素。

钢铁材料的热处理包括以下几个重要的操作工艺，它们习惯上被称为：正火、球化处理、去应力退火、退火、淬火、回火和表面硬化（渗碳）。

正火， 指把材料加热到临界温度以上 $100 \sim 200$℉ （$55 \sim 100$℃），然后在空气中冷却。正火温度比淬火温度高约 100℉ （55℃）。

正火的目的是改善钢铁材料在锻造中变粗的晶粒结构。对于大多数中碳锻钢来说，不论是否合金化，在锻造后和机械加工前通常都采用正火处理工艺，以有利于形成较均匀的组织，并且在大多数情况下还可以改善材料的切削加工性能。

高合金正火钢（气冷钢）不需要正火处理，因为正火处理后将使材料硬化而失去原有的性能。

球化退火 是退火的一种形式，在钢的加热与冷却工艺过程中产生圆形或球状的碳化物——钢中的硬质组成部分。

工具钢的球化处理是为了提高其可切削性能。球化处理是把碳钢加热到 $1380 \sim 1400$℉ （$749 \sim 760$℃），对于合金工具钢则更高一些，保温 $1 \sim 4$h，然后在热处理炉中缓慢冷却。

去应力退火， 这种方法用于去除钢成形、冷加工、焊接或机械加工过程中产生的内应力。这是最简单的热处理方法，把钢铁材料加热到 $1200 \sim 1350$℉ （$649 \sim 732$℃）后，在空气或加热炉中冷却。

大型模具通常是先进行粗加工，去应力退火后再进行精加工。这样在机械加工和热处理后产生的变形最小。由于不同的加热和冷却循环相结合以及截面的变化，在焊接截面将产生应力，此种应力在机械加工中将会引起相当大的变形。

退火，此种工艺是将钢铁材料加热到很高的温度后，保温一段时间，然后缓慢冷却。材料退火后将使组织均匀，并且建立与其特性相符的平衡组织。

工具钢购买时的状态一般是退火后。有时有必要对硬化后的钢再次加工，则必须对其进行退火处理。这种类型的钢进行退火时，钢被加热到略高于临界温度范围以上某个温度后再缓慢冷却。

淬火，此种工艺是将材料加热到临界温度范围以上，然后以足够快的冷却速度通过临界温度范围，使钢明显硬化。

回火，此种工艺是将淬火后的钢或合金加热到临界温度以下某个温度，以减少钢在淬火中产生的内应力。

表面硬化，在钢质零件的表面渗碳，然后对其进行淬火处理，它是热处理的一种重要方法。这种工艺包括采用熔融氰化钠混合物的液体渗碳，采用活性固体材料如木炭、焦炭、燃气及油等其他固体渗碳。

第 3 课　第三角投影法

六视图。任何物体都可以从六个相互垂直的方向进行观察，如图 3-1a 所示。如果需要的话，还可画出这六个视图，见图 3-1b。图示六个视图以美国标准排列，且俯视图、主视图和仰视图垂直排列，而后视图、左视图、主视图和右视图水平排列。将视图绘制在不适当的位置是一个严重的错误，而且常常被认为是绘图过程中可能出现的最为严重的错误之一。

注意，高度尺寸通常是在后视图、左侧视图、主视图和右侧视图上表现出来的；宽度尺寸通常是在后视图、俯视图、主视图和仰视图中表现出来的；而深度尺寸则是在围绕主视图的四个视图——左侧视图、俯视图、右侧视图和仰视图中表现出来的，且在每一个视图中，都表现出了两个主要的尺寸。还需要注意的是在围绕主视图的四个视图中，物体的前面（正面）正对主视方向。

相邻视图是相对应的，如果设想物体的主视图为图 3-1 中的主视图，从主视图的右侧看，得到的是右侧视图，用箭头 RS 表示。同样，如果设想右侧视图是物体的主视图，通过从右侧视图的左边看，则得到图示中的主视图，并用箭头 F 表示。同样的关系在任何相邻的视图中都存在。

必需的视图。用于生产中的图样，应当只包括能够对物体形状清楚、完整描述的视图。最低限度所需要的视图称为必需的视图。在选择视图时，绘图者应该选择能够最佳表示物体基本轮廓和形状，而且图中的隐藏线（虚线）最少的视图。

如图 3-1 所示，物体的三个不同特征需要在视图中表示：①从前面看，是圆形的顶和孔；②从顶上看，是矩形槽和圆角；③从侧面看，是带有内圆角的直角。

物体的三个主要尺寸是宽、高和深。不论物体的形状如何，在技术制图中，这些固定术语常常用来表示这些方向的尺寸。不用"长"、"厚"术语，因为它们不适应所有的情况。俯视图、主视图、右侧视图紧密排在一起，如图 3-1 所示，因为它们是最常用的视图，所以称为三视图。

视图的排列。视图排列中的错误是学生经常犯的，有必要再次强调。依据美国标准，所绘视图必须按图 3-1 所示关系排列。图 3-2a 所示为需要三视图表示的一个阶梯形零件，其三视图的正确排列如图 3-2b 所示。俯视图必须与主视图的两边对齐——而不能不成一直线，如图 3-2c 所示。即使视图与主视图在一条线上，但也不能将视图位置放反，如仰视图在主视图之上或右侧视图在主视图的左边（图 3-2d）。

第 4 课　公　　差

可互换性制造允许将零件分散在不同的地方进行制造，然后集中到一起进行装配。这些零件全部完全合适地装配在一起，是大批量生产的最基本要素。如果没有可互换性制造，现代工业就不可能存在；而如果没有工程师对零件尺寸的有效控制，可互换制造也就不可能实现。

然而，将任意零件制造得非常精确几乎是不可能的。零件可以制造成很接近要求的尺寸，甚至达到百万分之几英寸或千分之几毫米的精度，但要达到这样的精度是极为昂贵的。

在实际生产中，幸运的是大部分零件往往不需要非常精确的尺寸，这主要依据零件的功能要求而定。如果童车制造商将童车制造成与喷气式发动机一样的精度，制造商将会很快退出童车市场，因为没有人情愿支付昂贵的价格购买童车。因此，需要的是确定具体尺寸所要达到的精度的方法。这个问题的答案就是零件每个尺寸的公差技术要求。

公差是具体尺寸允许变化的总量，是最大和最小极限尺寸的差值。公差有两种表达形式：单边公差和双边公差。单边公差中，尺寸的变化完全在一边，如 "$30_{-0.02}^{0}$" 就是一个单边公差，这里名义尺寸 "30" 允许在 30mm 和 29.98mm 之间变化。双边公差中，尺寸变化是在两边，如 "30.00 ± 0.01" 或 "$30_{-0.10}^{+0.05}$"。在双边公差的第一种表达形式中，公差的变化比单边要均匀，尺寸变化是从 30.01mm 到 29.99mm；而另外一种形式中，尺寸的正负偏差可不同，这时尺寸变化是从 30.05mm 到 29.90mm。

在工程中，一般设计的产品是由许多零件组成的，并且这些零件相互之间以某些方式紧密配合，则在装配过程中，认真考虑两个零件间配合的类型是非常重要的，这实际上规定了两个零件在装配期间的装配方法。

以轴和孔的配合为例，在最简单的情况下，如果轴的尺寸小于孔的尺寸，孔与轴间存在间隙，称为间隙配合。另一种情况，如果轴的尺寸大于孔的尺寸，这种配合称为过盈配合。然而，由于尺寸的可能性，在实际装配过程中有时是间隙，有时是过盈，这种配合称为过渡配合。这些配合形式如图 4-1 所示。

第 2 单元

第 5 课　刀具设计

金属切削原理为我们检验刀具设计所涉及的所有要素提供了一个理论框架。我们需要将软得像黄油似的工件材料变得坚硬，并且使其具有良好的抗剪切性能。每种工件材料必须区别对待，随着工件特性之间差别的扩大，可以应用于每种工件材料的基本信息的数量却在减少，这个差别不仅指工件材料的不同，而且还指刀具形状及刀具组成的不同。

刀具设计者必须根据许多变化的因素全面协调考虑，以提供尽可能好的切削几何形状。过去人们通过大量的反复试验才能确定一个方案，但是如今随着刀具种类的不断增加，反复试验的成本太高了。

设计者必须不断增加自己的专业知识，主要是积累应用方面的数据，以及比较其他人的基本经验。如刀具制造者和材料销售商在他们公司发展后会拥有一个公司形象，形象意味着采购指南；而仔细搜索已有的文献和资料，将提供一个良好的开端，这比采用反复试验更简便。

采用机械加工的方法去除材料包括五个相互影响的要素：切削刀具、刀具夹紧和导向装置、工件夹紧装置、工件、加工设备。刀具可能是单刃或多切削刃，刀具可设计成直线运动或旋转运动。刀具的几何形状取决于它的功能。刀具的夹紧装置可用来也可以不用来进行导向和固定。刀具夹紧装置的选用取决于刀具的设计及功能。

实际工件的组成极大地影响加工方法、刀具组成与几何形状，以及切削速度的选择。工件的形状决定了机械加工方法及刀具运动轨迹（直线或旋转）的选择。对工件夹具的要求，在很大程度上是由工件材料的成分及几何形状决定的。工件夹紧装置的选择，还取决于夹具能对工件施加力的大小。刀具的导向属于工件夹紧装置的功能。

要成功设计出符合材料加工过程要求的刀具，首先必须完全了解刀具的功能和形状。这些知识能帮助设计者为完成某项任务选择正确的刀具。进而，刀具又决定了刀具夹紧装置及导向方法的选择，刀具的切削力又决定了工件夹紧装置的选择。尽管整个加工过程包括了五个相互影响的因素，但是它们都开始并基于一点，那就是刀具与工件接触点上所发生的情况。

制造一个特定形状及尺寸的工件，其主要方法就是用刀具的切削刃将多余的材料从工件上切削掉。尺寸比较大的原材料被逐步加工成所需要的特定形状，将多余材料从工件上切削掉的过程，通常称为机械加工成形或简称为机械加工。

成形一定的形状和尺寸，也可以通过采用热挤压或冷挤压、砂型铸造、压力铸造、精密铸造等方法来实现。通过外力可以成形或拉深金属板材。除了机械加工外，还可以通过化学或电的方法来完成金属的去除过程。种类繁多的工件即使不使用机械加工的方法也可以制造出来。但是从经济的角度来考虑，通常还是要求用机械加工的方法制造产品，或全部用机械加工或机械加工与其他加工工艺结合来完成产品的生产。

刀具都设计有锋利的切削刃，以使刀具和工件的切削接触面最小。切削刀具的形状变化影响刀具的寿命、工件的表面粗糙度以及切削金属所需的总作用力。刀具上不同的角度构成了刀具的几何形状。刀具的命名是以希腊字母"α"和一些数字构成的一个有序排列，其中的数字代表着一些刀具的角度、重要尺寸、特殊性能以及刀尖半径的大小。美国国家标准协会已将这种表示方法进行了规范，以碳钢及高速钢刀具为例，图5-1同时列出了刀具命名的基本要素。

第 6 课　工件夹紧原理

工件夹紧装置是包括保持、控制或夹紧工件完成加工过程的所有装置。夹紧力可由机械、电气、液压或气动装置提供。夹紧装置主要用于去除材料的加工过程，在机械加工中夹具是最重要的装置之一。

图 6-1 所示为材料成形加工中所有的基本组成元素的示意图。右手相当于刀具夹紧装置，左手相当于工件夹紧装置，刀子是刀具，木头则是工件。两手结合操作将木头切削成所需要的形状，并产生木屑。手所属的有机整体人可被看作施加力、运动、位移和控制其他零件的机器。除了增加力的元件外，这些基本元件存在于所有应用刀具夹紧装置和工件夹紧装

置的加工中。

图 6-2 所示为在研磨加工中所使用的夹紧圆棒的夹钳（手钳）。这种较简单的夹具图解释了用杠杆原理增加力的方式，并且也说明夹紧圆棒的接触位置采用锯齿结构以增加防滑性。

图 6-3 所示为普遍使用的螺杆式夹钳（台虎钳）。丝杠推动活动钳口运动提供夹紧力。台虎钳是通过螺旋的自锁特性来锁紧的，它提供了使其他部件安装到机床上的一种手段，并确保加工时的准确定位。

用于夹紧装置上的各种钳口如图 6-4 所示。由液压装置施加夹紧力，而丝杠的作用则是使钳口接触工件，钳口可以做成如图所示的凹进的形状以固定特殊形状的工件。此外，许多复杂的工件需要复杂的钳口形状相匹配。

另一种形式的夹紧装置是卡盘。在车削、镗削、钻削、磨削或其他回转加工中，所用到的各种机床卡盘用来装夹刀具或固定工件。卡盘产品有多种规格。一些是通过扳手手工夹紧，其他则是通过气动、液压或电动装置来夹紧固定。某些卡盘的夹头是独立进给和固定的，而另一些卡盘则是所有的夹头统一进给的。图 6-5 为通过四爪单动卡盘夹紧工件。钻削工件的钻头夹持在万能卡盘中。

夹紧装置的目的和作用。 当夹紧装置夹紧工件时，必须使工件与刀具保持正确的相对位置，并且在加工过程中，在夹紧力和切削力作用下能保持精确的位置。夹具由几个基本元件组成，每个元件都具有一定的功能。定位元件起固定工件位置的作用，结构元件或夹具体承受各种力的作用，支架使夹紧装置固定在机床上。夹钳、丝杠、卡盘提供夹紧力，各种元件可以通过手动或机械操作。夹紧装置的功能是夹紧可靠，定位准确，保证操作者和机器设备高度安全。

夹紧装置的设计与选择受许多因素的限制，最重要的是夹紧装置本身的物理特性。夹紧装置需要有足够的强度支撑固定工件，使其不产生偏移。在考虑工件材料的前提下，必须仔细选择夹具的材料，只有这样才不会引起接触性破坏。例如：若选用硬的钢质台虎钳口，就会使比较软的铜质工件材料表面受到破坏。

在机械加工过程中，切削力的大小和方向是变化的。钻孔可以产生扭矩，刨削加工可以导致直线式推力。夹紧装置则为工件提供反作用力；对于一些特殊的机械加工，也需要设计夹紧装置。

许多在工业中使用的夹紧装置，不仅仅是用在材料去除加工过程中，还可用在工件检验、装配和焊接等工艺过程中，各种夹紧装置在结构设计与外观上会有一些小的差别。一些通用夹具既可以用在车削加工中，也可用于同一或其他工件的检验中。

第 7 课　钻模和夹具设计

钻模是用来夹紧、定位、支撑工件并且对刀具进行导向以通过它完成切削循环的夹紧装置。钻模有两种类型：钻削夹具和镗削夹具（译者注：jig 可以译为"夹具"，但从上下文看，这里单指钻削夹具和镗削夹具，其他夹具则用 fixture 表示）。钻削夹具是最通用的夹具，在钻孔加工、螺纹加工、铰孔加工中使用，也可在锥孔加工、扩孔加工、倒角和锪孔加工中使用。另一方面，镗削夹具用于对已有孔进行高精度的加工中。两种类型的夹具结构基本类似，唯一的不同是在机械加工中镗削夹具装有导向套以固定镗刀伸出端的刀杆。

设计钻模时，必须考虑到各种因素的影响。虽然对定位元件、支撑元件、夹紧元件已经作了介绍，但本节仍需阐述这些要素，因为它们应用于夹具设计。由于所有的夹具具有类似的结构，因此某种夹具需要考虑的因素也适用于其他类型的夹具。在夹具的设计与选择时，首先需要对工件和加工方法进行认真分析。

在设计任何夹具时，首先要考虑的问题是制造该工具的成本和使用该工具进行生产所希望产生的效益两者之间应保持相对平衡。节约的生产成本应大于夹具的设计与加工成本。大多数情况下，工具设计者必须仔细估计专用夹具的成本，包括熟悉零件图、工艺规程和相关文件。

如果将成本作为专用夹具的设计依据，还必须考虑产品的复杂程度、加工孔的数量与位置、所要求的尺寸精度及所要加工零件的批量等。一旦夹具设计者认为专用夹具的成本适合生产加工，就需要分析和搜集资料设计合适的夹具。

在机械加工中，夹具是用来夹紧、定位和支撑工件的。与钻（镗）模不同，夹具对刀具没有导向作用，但它却为刀具对准工件提供了参考。夹具是按其所使用的机床进行分类的，并可对夹紧装置进行细分，即按机械加工的具体成形方式细分。例如：铣床所用的夹具称为铣床夹具；然而，如果成形加工方式是组合铣削，夹具就称为组合铣削夹具；同样，为切槽所设计的带锯夹具称为带锯切槽夹具。

钻模和夹具都需要设计夹具体。就大多数情况而言，夹具在设计上比钻模能够承受更大的应力，并且总能可靠地将工件夹紧到机床上。由于这些原因，设计者必须要注意夹紧任意零件时，定位、支撑和夹紧的方法。

如果想设计一副好的夹具，还需要考虑一些其他因素，如成本、生产能力、加工工艺、工具寿命等因素需要和工件一样受到重视。

夹具设计过程中，首先考虑的是夹具成本与采用夹具生产所能获得利润的关系。生产批量、生产率和零件精度必须作为专用夹具增加成本的依据。夹具的设计成本必须以尽可能少的加工时间和节约产品的成本作为补偿。

第 3 单元

第 8 课　压力机的类型

压力加工的特点是每隔很短的时间，压力机通过模具在工件材料上施加压力，从而使材料切断或成形。

压力加工是指通过压力来完成的，通常在很短的时间内加工出精度较高的零件。

压力加工中的工作压力是由压力机产生、并对其进行导向和控制的。

动力压力机。压力机由机身、工作台、垫板和往复运动的滑块组成。通过装在滑块和工作台间的模具对材料施加压力。

储存在压力机旋转飞轮中的能量（液压压力机由液压传动装置提供）转化为滑块的往复直线运动。

压力机的类型。开式双柱可倾压力机（见图 8-1）也称为开式压力机，装有一个 C 形机架，从这里可以接近工作台至滑块间的工作空间。机架相对于水平面可倾斜一定的角度，并且使成形零件在重力作用下下落。这种开式后倾式压力机允许坯料、工件、成品从前至后地进给或卸料。

压力机的主要部件包括：

1. **机身**。它是在中心开有口的矩形框架，用来支撑工作台。

2. **工作台**。它是一个2~5in（51~127mm）厚的平钢板，冲压模具及附件安装在工作台上。具有标准尺寸和开口的工作台可以从压力机制造商处获得。

3. **滑块**。滑块在压力机上半部分，它在一个行程中移动的距离由压力机的设计尺寸决定。滑块的位置可以进行调整，行程不可调节。

从工作台的上表面到滑块的下表面之间的距离可以调节，称为压力机的闭合高度。

4. **打料装置**。它是压力机向上行程中的一个机构，它从压力机上顶出工件或废料。

5. **气垫**。装在压力机工作台下或内部的附件，可以产生向上的运动或力，由空气、液压、橡胶、弹簧驱动。

传统的闭式压力机在机身后部有一立柱。一般具有方形或矩形进料口，允许条料、工件或成形零件进给或卸料。

由于特殊的用途，这种类型的压力机也可以从前向后进料，如图8-2所示。

板料折弯机除了机身比较大外，多为6~20ft（1.8~6m）或更大，基本上类似于开式压力机。板料折弯机一般用于大型金属板料的弯曲操作，也可以通过冲压模具实现一系列冲孔、切边、成形等分离操作。板料折弯机可以加工复杂的零件，而不需要高成本的冲压模具，这主要通过把复杂的零件操作工艺分成几个简单的操作来实现。此种类型的操作常常用在小批量生产或样品试制中，模具成本一般比较低，但是人力成本相对比较高。由于零件的传递与定位在每一个位置是由人工操作的，操作者还必须时刻遵守安全操作规程，以避免受到伤害。

液压机主要用于成形操作，与多数机械式压力机相比其操作的工作循环慢。它的优点是工作压力、行程和滑块的速度可调，如图8-3所示。

双动压力机主要用于较大型金属零件成形或深拉深成形，这种压力机有内滑块和外滑块。在操作过程中，外滑块首先与工件接触并且压紧，然后内滑块对工件进行拉深成形。

三动压力机具有和双动压力机相同的内、外滑块。此外，三动压力机工作缸还有另一个滑块，它可向上运动，从而在一个冲压循环中允许反向拉深。这种压力机的应用并不广泛。

肘杆压力机用于精密冲压及压印，驱动的设计允许在滑块行程的底部有非常高的冲压力，这种压力机采用曲柄使两肘杆连接处围绕运动死点中心来回摇动，因而滑块行程短，在接近行程底部时运动有力而速度缓慢。

压力机有很多种类和特殊的用途，以上介绍的是在工业方面应用较多的几种压力机。

在压力机的设计方面，操作者的安全也是必须注意的一个基本问题。在滑块下方放置的安全滑块必须能阻止滑块的惯性下落，滑块须锁定在一个固定的位置。在压力机运行期间，必须采取适当的防护和遵循安全操作程序。

根据OHSA（职业健康与安全法令），压力机安全也是法令，必须严格遵守安全规则。

第9课　注射成型机

大量的塑料制品是采用注射成型工艺生产的。注射过程（工艺）包括两个阶段：一是经料斗送入的粉末状或粒状形式的塑料混合物通过熔融区和定量区；二是将熔融塑料注射到模具型腔中。经过短暂的冷却阶段后，打开模具，固化的塑料零件被推出。在大多数情况下，这些零件就可以直接使用了。

将熔融的塑料注射到模具型腔的形式有很多种，在大型注射机里最常用的形式是往复螺杆式注射装置，如图9-1所示。

螺杆集注射装置与塑化装置的作用于一体，当物料加入到旋转的螺杆中，物料依次经过三个区域，分别为加料区、压缩区、定量区。在加料段之后，螺杆的螺槽深度将逐渐变浅，从而对塑料产生压实力，并通过对塑料的剪切作用转化为加热塑料的热能，使塑料成为黏流态。在定量段，料筒表面通过热传导提供额外的热能。当螺杆前部的加料室被填实后，将会迫使螺杆后退，并断开行程限位开关而开动液压缸，从而使螺杆向前运动，并将熔融态塑料注射到闭合的模具型腔中。使用一个止逆阀可以阻止塑料在压力作用下倒流回螺旋槽。

注射机所能提供的锁模力是尺寸设计的一部分，以吨力为单位。对于一个特定的工作过程所需要的吨位可以用经验方法来确定，锁模力一般取 $2tf/in^2$ （30.4MPa）。如果塑料流动困难或塑料制件是薄壁件时，可以选择 $3 \sim 4tf/in^2$ （45.6 ~ 60.8MPa） 的锁模力。

许多往复螺杆式注射机也可以用来成型热固性塑料。很显然，这种塑料通常采用压缩或传递模塑的成型方法。热固性塑料在模腔内固化或者发生聚合反应后取出，其温度范围为 $375 \sim 410°F$ （190 ~ 210℃）。热塑性塑件必须在模具内冷却到一定的温度，以防止在取出时发生变形。这样看来，热固性塑料的注射成型周期较短，其模具必须被加热，而热塑性塑料成型时则需要冷却。

图9-2所示为注射模塑的几种方法。最早的是单级柱塞法，当柱塞后退时，物料从料斗进入加料室，然后柱塞向前挤压物料经过加热室，物料在加热室中被软化并最后在压力作用下射入模具。目前单级往复螺杆式注射装置应用广泛，因为它塑化物料更彻底，而且塑化效率也较高。当螺杆旋转时，将受到向后的推力，压实从料斗进入加热室的物料。当加入足够多的物料后，螺杆停止旋转，并开始像柱塞一样向前驱动以压实物料。在两级预塑式注射装置中，物料是在一个预塑料筒中塑化，柱塞或螺杆的运动将一定量的塑料送入注射室，然后柱塞将塑料从注射室注入模具中。

一台注射成型机在一个注射周期中完成对热塑性塑料的加热软化、成型和冷却硬化等过程。工艺温度通常为 150 ~ 380℃ （300 ~ 700°F），注射压力通常取 35 ~ 350MPa （5000 ~ 500000lb/in^2）。模具采用水冷方式冷却。当打开模具时，成型的塑件和浇注系统凝料从注射端退出，并从另一侧被顶出。然后模具重新闭合并锁紧，开始进入下一个工作周期。热固性塑料可以注射成型，但必须在下次开启设备之前聚合和成型，这可以在往复螺杆式注射机上完成，注射时每次加入的物料均需加热至固化温度。通过另一种方法（也叫喷射成型法），热固性塑料可以在单级柱塞式注射机内成型。

这些设备也可以用来成型分层的塑料制品。一套液压缸和柱塞将一定量的表层塑料注入模具，然后另一个液压缸喷射助剂。最后，从第一个液压缸进行的最后喷射将心部的塑料从浇口处分开，其目的就是生产具有最佳性能的分层复合材料，心部和表层的塑料均可发泡成型。

第 4 单元

第 10 课　冲裁工艺

冲裁操作。 在以下的讨论中，将会经常使用一些模具专业术语。图 10-1 所示为一些常见的术语。

冲裁操作中的剪切动作。 金属材料在冲裁模具间的冲裁切断是剪切应力达到断裂点或超出材料强度极限的一种剪切过程。

金属材料承受拉应力和压应力，如图 10-2 所示。在超出弹性极限时材料将会伸长，然后开始塑性变形，断面减小，最终裂纹沿不断减小的断面上的撕裂带扩展，从而实现完全分离。

图 10-3 描述了冲裁和剪切的基本过程。处于金属上部的凸模施压使金属变形并进入凹模入口。当继续加载超出材料的弹性极限时，金属的一部分将被压入凹模入口，在材料下表面形成压痕，上表面也相应地发生变形，如图 10-3a 所示。当载荷继续增加时，凸模将把金属压入到某一个深度，直至压入凹模的深度等于金属的厚度，如图 10-3b 所示。这个压入过程发生在剪切引起的断裂开始和金属横截面减小之前，在上、下剪切刃处缩小的断面上均出现裂纹，如图 10-3c 所示。对于被剪切的材料，若间隙适当，则裂纹将相向扩展并最终相遇，从而引起材料的完全分离。凸模继续下行将冲下的部分通过料架推入凹模的漏料孔。

压力中心。 如果所冲裁的零件轮廓是不规则形状，滑块一侧的剪切力的总和将远远超过另一侧的剪切力。这种不规则形状将导致压力机滑块处在弯曲力矩的作用下，易产生偏斜和变形。因此，寻找一个能使剪切力的总和对称（平衡）的点是很有必要的，这个点就被称为压力中心，它是冲裁轮廓线的重心，而非（封闭区域内）面积的重心。

设计冲压模具时，尽量使压力中心位于压力机滑块的中心线上。

间隙。 间隙是一副模具凸、凹模之间的配合空间。剪切刃之间的合理间隙可以确保裂纹汇合。剪切刃的撕裂部分将形成一条光亮带。为了保证剪切刃的最佳精度，合理的间隙是必要的，其大小取决于所冲压成形材料的种类、厚度、韧度（硬度）。图 10-4 所示为间隙、塌角深度及断口。图 10-5 描述了毛坯或落料件上的断面特征以及合理的间隙。毛坯的剪切断面的上部拐角与落料件的下部拐角都将形成一个圆角，在此恰好是凸、凹模刃口接触材料的区域。这个圆角是由于材料发生塑性变形引起的，并且在冲裁比较软的金属材料时圆角会更明显。过大的间隙将在这些拐角处形成更大的圆角，而在相对的另一面的拐角处出现毛刺。

第 11 课　冲孔模和落料模设计

冲孔模设计。 在一个压力机行程中，在原材料上冲裁出两个孔的一个完整的冲压模具如图 11-1 所示。一家大制造商将其分类并标准化为单工位冲孔模。

任意一副完整的冲压模具，都是由一对（或多对）生产冲压零件的配合零件组成的，同时还包括模具上的固定支承零件和传力连接零件。压力加工术语上通常把所有冲压模具上的凹件（母件）定义为凹模。

导柱安装在下模座上。上模座安装与导柱滑动配合的导套。上、下模座连同导柱、导套组合成的部件称为模架。市场上有多种模架尺寸和设计形式供选用。如图 11-2、图 11-3 所示，导柱沿卸料板的垂直方向导向，在图 11-3 的剖视图中导柱没有表示出来。

安装在上模座上的凸模固定板固定了两个圆形凸模（模具中的凸出部分），这两个圆形凸模通过嵌入在卸料板上的模套进行导向。安装在一个细小凸模外围的护套可以防止凸模在压力机滑块的压力作用下弯曲。在冲裁完工件以后，两个凸模还要伸入凹模衬套一小段距离。

凹模由两个嵌入在凹模板内的衬套组成。由于冲孔模应冲出所要求的孔径，所以凹模的衬套内径应稍大于凸模一个间隙值。

当废料或工件黏在上行的凸模上时，必须将其从凸模上卸下。弹簧驱动的卸料板相对凹模板运动，将原材料卸下，直到凸模从冲过的孔内抽出。被冲孔的工件用定位板定位，如图 11-1 所示，定位板是与冲件外形轮廓相符的平板。在滑块下行之前，毛坯要在凹模内通过定位销、定位板或其他类型的定位装置定位。

落料模设计。除了用一块凹模板代替了凹模衬套和用一个落料凸模代替两个冲孔凸模，而坯料的定位用挡料板代替外，图 11-2 所示的小型落料模具的设计与图 11-1 中的冲孔模具设计相类似。由于冲出的落料件经凹模、下模座和压力机垫板内孔下落，所以这是一副下出料方式的结构设计。

大型落料件通常采用倒装式落料模（见图 11-3），其凹模被安装在上模座，而凸模被固定在下模座。大型落料件通过垫板的下出料方式来实现通常是不切实际的，而采用镶拼式结构设计则比较合适（见图 11-4）。

对于倒装式落料模，（漏料孔）斜度或倾斜角处的间隙是不必要的，因为落料件不会采用下出料方式。为了使冲裁刃口结构简单，便于修磨，保证强度，每一个拼块不应该存在刃口轮廓线的交点和复杂轮廓。图 11-4 所示的模具中拼块 1 和拼块 2 包含了整个半圆轮廓，而其他 6 个拼块包含了所有的直线轮廓。

弹性卸料板固定在下模座，它向上运动进而从固定在下模座上的凸模上卸掉废料，卸料螺栓对卸料板起固定和导向作用。

压力机滑块回程时，打杆的上端将撞击压力机横梁，打杆受力向下运动，其下端通过凹模将冲制的零件从模腔内推出。限位环用来挡住打杆并限制其运动行程。

第 12 课　复合模和级进模设计

复合模设计。在压力机的一次单行程中，复合模只能完成冲裁工艺（通常为冲孔和落料）。复合模可以生产平面度和尺寸公差精度较高的冲孔落料件。复合模的特征是落料凸模与落料凹模是倒装的。如图 12-1 所示，落料凹模被固定在上模座，而落料凸模则被固定在下模座。落料凸模同时也起到冲孔凹模的作用，在下模座及冲孔凹模内有一锥形孔作为漏料孔。

在压力机滑块的上行行程（回程）中，压力机的打杆碰到打料环，作用在打杆上的力使卸料装置下移，这样就会将成品件从落料凹模中推出。条料由固定在弹性卸料装置上的导料板导向。当卸料板向上运动时，将使废料从落料凸模上卸下。在冲裁操作开始前，卸料板与落料凹模的下平面将板料压平。

四个特殊的带肩螺钉（卸料螺钉）在工业上是能用零件。该螺钉在卸料板的卸料行程中起导向和固定预压弹簧的作用。

落料凹模与凸模衬套用螺纹连接，并用销子固定在上模座上。

当弹性推件杆从凹模内推出落料件时，在推件块上装的弹性顶料销（油封开关）被压下。在滑块上行时，顶料销击破落料件和顶料销的表面油封，使零件从落料凹模内落下。

级进模设计。级进模是指在压力机每一次行程中，在两个或更多的工位上同时完成一系列的金属板料冲压工序操作，其目的是当金属板料通过模具时能够生产出完整的工件。这种模具又称为拖件前进的连续模、系列压模或复合模。每个工位完成一个或更多不同性质模具的功能操作，但条料的移动必须从第一个工位起通过随后的每一个工位，以便生产出完整的工件。模具上的一个或更多空位可以合并，这些工位虽然不能直接加工金属板料，但它们具有定位条料、推动条料在工位中间的移动、提供最大尺寸的模具截面、简化模具结构等作用。

在压力机的每一次行程中，金属条料的线性移动被称为连续进料或进给，即节距，并且节距等于每个工位中间的距离。

需要多个工序操作才能完成的零件，可以在高生产率的级进模上实现。金属条料可以卷料的形式进给，并且不同的工序操作（如冲裁、落料、切口）可以在压力机的同一个工位上通过每次行程中的多个凸模冲头来完成（见图12-2）。

第13课　卸料板和打料装置

卸料板。卸料板有两种类型：固定式和弹性卸料板。任何一种卸料板的主要功能是从冲裁或非冲裁凸模和凹模中卸下工件。卸料板迫使工件或废料从模具中脱离，也称为打料，装在内侧的卸料板，也叫顶件装置。除了基本功能外，卸料板还具有压料即夹紧作用，定位、导向板料和工件的作用。

卸料板通常是与凹模板的宽度和长度尺寸相同。在简易模具中，卸料板可以用同样的螺钉和暗销固定在凹模板上，并且螺钉头要沉入到卸料板中。在比较复杂的模具或者是由可拆卸的部件拼制成的模具中，凹模板的螺钉通常是倒置的，并且卸料板的紧固是独立的。

卸料板的厚度必须足以经受从凸模上卸下条料所需的力，再加上从条料经过卸料通道所需要的其他力。除了重型模具或具有大的落料截面外，对于沉孔螺钉，所需要的厚度范围通常为3/8～5/8英寸（9.5～16mm）就能满足要求。

条料经过卸料通道的高度将至少保持在条料厚度的1.5倍。如果条料抬起超过固定挡料销，其高度尺寸还要适当增加。卸料通道的宽度应该是条料的宽度再加上适当的间隙，以利于条料切断的宽度可以有所不同。

选择使用弹性卸料板的方法取决于所需要的压力大小、空间限制、模具形状、工件的特性和生产设备等。图13-1所示为一些弹性卸料板的应用实例。

打料装置。因为切断后的坯料由于摩擦的作用而滞留在凹模内，因此，必须要在冲头上升行程时从凹模内推出坯料。打料装置由推板、推杆和保持环组成。推板与凹模口内轮廓为间隙配合，当坯料被切断后能够向上运动。通常用铆钉将一个较重的推杆连接到推板上，而且推杆可以在模座的模柄孔内滑动。推杆可伸出模柄以上，并且通过保持环来保持和限制打料装置的行程。在接近冲头行程的上限时，压力机上的打杆将与推杆撞击，从而顶出坯料。

因为有些危险在偶然情况下可能会发生，因此，保证打料装置的可靠性是很重要的。

在零件的推出过程中，以及在零件形状和模具选取允许使用的情况下，正向打料装置与弹性卸料板相比有下列优点：

1. 零件处理自动化。在接近冲头行程顶部时，顶出的坯料能够被吹向压力机的后部，或者也可以通过倾斜压力机获得同样的结果。

2. 模具成本低。一般情况下，打料装置比弹性卸料板的制造成本要低。

3. 直接作用。打料装置不会像弹性卸料板有时会发生粘连现象。

4. 压力要求低。在冲头下降期间，没有大型弹簧需要进行压缩。

图 13-2 所示为几种设计相对较好的打料装置。

如图 13-2a 所示，这个设计很简单，被应用到一个普通的倒置式复合模。它由一个伺服撞杆 1，打料板 2，和用暗销连接到撞杆 1 上的停止环 3 组成。卸料装置 4 由一个借助固定螺钉限制其行程的弹簧返回支承销组成。

如图 13-2b 所示，打料板也被用作导向细长凸模通过淬火导套的工具。

如图 13-2c 所示模具设计，冲压后法兰壳在上模中向上运动并且通过正向打料装置被顶出。

第 14 课　弯曲模设计

弯曲是沿平板料或条料位于中性面内的中心线（通常为沿长度方向）发生的均匀变形。由于金属流动发生在金属的塑性变形范围内，因此当外加载荷去除后将会保留一个永久性的弯曲变形。弯曲件内表面呈现压缩变形，外表面呈现伸长变形。普通弯曲变形不可能在被弯曲的金属材料上准确地再现凸、凹模的形状，凸、凹模形状再现是成形工序中的一种。中性层是在弯曲材料上变形为零的平面。

弯曲半径。最小弯曲半径因材料而异，通常情况下，退火状态的金属材料能被弯曲到与材料厚度相等的半径而不会发生弯裂现象。

弯曲件的展开长度。由于弯曲后被弯金属会变长，一般产品设计者应考虑其增加的长度，如果弯曲件的长度公差非常严格，那么对于模具设计者也必须考虑其长度的变化。弯曲金属的长度可用如下公式计算：

$$B = \frac{A}{360} 2\pi(R_i + Kt)$$

式中　B——弯曲件沿中性轴展开长度（mm）；

　　　A——弯曲中心角（°）；

　　　R_i——弯曲件内表面半径（mm）；

　　　t—— 材料厚度（mm）；

　　　K——0.33（$R_i < 2t$）或 0.5（$R_i > 2t$）。

弯曲方法。在弯曲模中通常采用两种弯曲方法。由 V 形块支撑的金属条料或板料（如图 14-1a 所示），在楔形凸模的作用下进入 V 形块内。这种弯曲方法称为 V 形弯曲，使弯曲件成形的弯曲角为锐角、钝角或直角。在 V 形模具内，弹性滚花压销和零件之间的摩擦将会阻碍或减少零件在弯曲过程中的边缘偏移。

折边弯曲是对横梁的悬臂部分的加载（见图 14-1b），弯曲凸模 1 施压于金属，使之背靠凹模块 3。弯曲的中心线平行于凹模边缘。在凸模接触坯料之前，坯料被弹性压料板 2 压紧在凹模块上，以防止在凸模下行时金属的窜动。V 形弯曲模和折边弯曲模如图 14-2 所示。

弯曲力。 V 形弯曲力计算公式如下：

$$P = \frac{KLSt^2}{W}$$

式中　P——弯曲力（t）（对于米制用法，乘 8.896 可转化为 kN）；

　　　K——开模行程系数，当开模行程是金属厚度的 16 倍时为 1.2，开模行程是金属厚度的 8 倍时为 1.33；

　　　L——工件长度（in）；

　　　S——极限抗拉强度（t/in²）；

　　　W——V 形或 U 形模具宽度（in）；

　　　t——金属板料厚度（in）。

U 形（槽形）弯曲力大致是 V 形弯曲力的两倍；折弯力为 V 形弯曲的 1/2。

回弹。 当作用在金属板料上的弯曲力去除后，弹性应力也随即消失，这将引起金属板料的移动，从而导致弯曲角减小（即弯曲部件间的包角增大）。这种金属板料的移动被称为回弹。钢材的回弹角在 0.5°~5°范围内变化，这主要取决于金属材料的硬度。磷青铜的回弹角可达 10°~15°。

V 形弯曲模的回弹补偿原则为 V 形块或楔形凸模的弯曲角小于零件需要的弯曲角。由于零件弯曲时比需要的弯曲角稍大，因此回弹后可得到需要的弯曲角。

采用其他类型的弯曲模生产弯曲制件时，可通过切槽或做成台阶凸模过量弯出一定的角度来补偿零件的回弹量。

第 5 单元

第 15 课　压铸模

多数压铸模是由铁质金属材料零部件组装而成的，而且每个零部件在机构中都起到一定的作用。熔融金属在一定压力下被注射入模具型腔后立即冷却，致使这些模具部件在温度快速变化的条件下工作。其中一些零件仅仅起到固定作用，而另外一些零件，例如模具镶块和型芯，它们必须能够承受熔融金属的冲击和高温。推出机构和侧抽芯机构必须在模具温度不停变化的状态下流畅地工作。还有一些零件起轴承导向作用，它们是由非铁金属材料（如磷青铜）制成，或是对零件表面进行渗氮或其他处理，以增强其耐磨性。

压铸时起成形作用的模具镶块必须进行精加工和热处理，目的是在高温环境下尽可能使镶块具有最好的力学性能。大多数压铸模的操作都是自动化，在没有延迟的情况下可以提高生产率，但若中途停止，则相应的成本增大。在过去的十几年中，从改进的电火花腐蚀到计算机控制的模具加工等尖端的加工方法，使模具的制造发生了变革。除此之外，科学研究委员会正在支持一个大型研究计划，包括模具热处理和表面涂层、计算机辅助模具设计研究，尤其是机械加工工艺和模具制造的经济性研究。

一套压铸模具可能包括十种或更多的不同种类的钢铁材料，还包括一些非铁金属和特殊耐热合金。图 15-1 和图 15-2 所示为压铸模典型零件示意图，每个零件都涉及设计、工艺和冶金问题。

低合金钢和铸铁零件。 推出装置组件由几个矩形板叠加组合而成，包括由圆柱形截面的推板导柱导向的推板。组件中的推板和推板导柱的材料是低碳钢（软钢），含碳量大约为0.15%（质量分数）。导柱和推出机构的限位装置还要进行表面硬化。推出装置中的零件起支承和导向作用，但不承受冲击载荷。支承板承受机械冲击和压力，但不承受强烈的热冲击，其材料通常是中碳钢。这些零件也可以用铸钢或球墨铸铁制成。

模板包括组成铸型型腔的凹模和型芯的运动机构，以便使模具能安装在压铸机的拉杆之间。模板的材料通常是中碳钢。例如英国标准中规定的 BS 970 08M40（En8），其成分质量分数为含碳0.4%，含锰0.8%，含硅0.3%。有时也用预硬化钢，其典型成分（质量分数）是 0.35% C，1.0% Mn，0.5% Si，1.65% Cr，0.5% Mo，这种成分在美国钢铁学会标准中为 P. 20。

模板被加工成有凹槽的形状，镶块就固定在模板的凹槽中。或者，也可用其他方法开槽，大型模板由中碳钢或铸铁制成，在铸造时就尽可能地铸成所需形状。模板的表面要做成平面，以便使射入的合金在相当大的注射压力下不会溢出。最好是将分型面设计在一个平面上，但有时将其设计成不规则形状，目的是防止产生侧凹。分型面和模板加工好后，将导柱（其直径根据模板的大小通常在 10~60mm 之间）固定在其中一个模板中（导柱固定在动模板，伸入定模板），以保证两个模板准确对位。在另一个模板中（定模板），加工出尺寸适当的孔，用来安装导套，导套通常由渗碳钢制成。在结构对称的模具中，其中一个导柱的位置是偏移的，目的是确保模具准确组装。导柱通常是用经过表面处理的低碳钢制成，有时也用经过表面处理的镍钢，以增加其强度。

如果位置允许，导柱安装在固定凸模的一侧（通常是动模或有推出装置的一侧），当模具从压铸机上卸下进行维修时对型腔起到保护作用。如果导柱这样安装会与复位杆产生冲突，尤其当导柱采用固定路径（开模方向）拔出，且不能提供横向运动时，导柱就需要重新定位安装在定模部分，如图 15-1 所示。方形导柱通常与大型模具一起用在锁模力在 600tf（约 6000kN）及以上的压铸机中。采用这样的模具，系统便于调节模具零件引起的热膨胀。

第 16 课　锻　　模

锻造正如我们所知是最古老的金属加工工艺。在人类文明的初期，人类就已经发现，加热后的金属很容易被锻打成不同的形状。锻造是通过锻打或冲压将金属加工成所需的形状，通常是在金属经加热塑性得到改善之后。在大多数情况下，需要将锻造的金属加热到标准的锻造温度，但有时也可进行冷锻。冷锻属于冷加工，是在室温到金属的临界温度之间完成的。

主要的锻造工艺。 在锻造过程中，可以对金属进行以下加工：

1. 拔长：增加金属的长度，减小其横截面积。
2. 镦粗：增大金属的横截面积，减小其长度。
3. 挤压：在闭合的型腔模中进行挤压。

优良的力学性能。 对于具有同样化学成分的锻件和铸件，锻件的力学性能一般比铸件好，所以，当零件必须承受危险应力时，最好采用锻造成形。锻造比铸造具有更好性能的主要原因有三个：第一，合理控制纤维流线方向可以提高强度；第二，锤击或压力机锻造过程可以得到致密的组织，通常能避免缩孔、气泡和缩松的产生；第三，锻造有助于细化金属的晶粒，锻造加工使金属沿结晶面产生滑移从而使粗晶粒细化。

锻造工艺的类型。锻造可按下面四种主要方法进行分类：

1. 锻工锻造。

2. 落锤锻造。

3. 压力机锻造。

4. 镦锻。

锻模设计。只有当锻件的最后形状确定以后，才能生产给定锻件所需的锻模。所以，对模具设计者来说掌握锻模设计的基本原理是非常重要的。模具设计时要考虑到以下几个方面：锻模斜度、分型面、倒角半径、收缩率和模具磨损量，以及模具错配、配合公差和加工余量。

在锻工锻造过程中，要用到带平面的一套模具和多功能手工工具。零件的成形取决于锻工的技术，即锻工移动金属、操纵锻造的过程。手工锻件的尺寸不精确，因此对表面尺寸要求高的锻件还要进行机械加工。这种锻造方法适用于小批量生产或预锻，预锻的锻件要通过其他锻造加工才能得到所需形状。这种方法加工锻件的重量在 1 磅（0.45kg）到 200 多吨（2×10^5 kg）之间，通常使用蒸汽-空气模锻锤、大型液压机、空气锤或者摇锤进行锻造。

锤锻模。锤锻模要求能承受极大的压力，耐磨，能将裂纹控制在最小尺寸，同时在大批量生产的情况下使用寿命长。为了能具备以上性能，模具材料采用铬-镍-钼合金钢、铬-镍合金钢或者铬-钼合金钢。确定锻件分型面时，必须考虑金属的流动性以及所产生的纤维流线的方向。如果可能的话，应该尽量选择水平方向上的平直分型面，因为不规则的分型面可能会造成压力（中心）的偏移，而且会增加模具的成本。标准拔模斜度是 7°，过小的拔模斜度会引起模具的快速磨损，增加锻件黏模的可能性。由于冷却时几股冷却液冲向锻模，锻件将沿着模具内表面收缩，所以内表面的拔模斜度要求大一些，通常是 10°。

无拔模斜度锻模。无拔模斜度锻造常用于锻造非铁金属及合金。模具通常安装在成套冲模中，在锻造压力机中工作。根据锻件的外形，模具型腔的两块或更多块是可以移动的，以提高锻造的通用性，但生率较低。在许多情况下，坯料第一次锻压时锻模没有完全闭合，多余的材料被切除，然后坯料被再次加热进行第二次锻压，此时模具完全闭合。

虽然这种锻造工艺费用比较昂贵，但二次机械加工的费用没有了，模具无须加工斜角，使锻件的总成本降低了一些。

第 6 单元

第 17 课　塑　　料

所谓塑性，是指材料是可塑的。根据定义，塑性材料具有在任何方向上连续变形而不发生断裂的能力。由此可得出塑性材料包括玻璃、金属及蜡等。因而，塑料这个名字容易引起误导，是不完全正确的。这里所讲的塑料专指合成树脂材料。合成树脂是经过各种化学工艺研制而成的。塑料制造业是当前相对流行的行业。1839 年，查尔斯·固特异发明了硬橡胶，1869 年，海亚特对赛璐珞的研制标志着塑料制品的出现。直到 1909 年，由 L. H. 贝克兰和他的同事研制并推广了酚醛树脂（目前最重要的材料之一）。今天，塑料广泛用于生产家用产品、汽车以及更多其他产品和机器。用电设备需要具有电绝缘性能的塑料零件。至今所生产的最大塑料产品之一是重 2600 磅（1179kg）、长 28 英尺（8.53m）的船身，它由玻璃纤维增强聚酯树脂经低压层叠而成。一些塑料的特殊性能被应用在很多特殊的场合，这些特殊性能的优点应该被尽可能地发挥。

热固性塑料和热塑性塑料。塑料材料通常叫作塑料，一般分为热固性塑料和热塑性塑料。经过初步的加热，热固性塑料软化和熔融，熔料在压力作用下充满模具型腔。在继续加热作用下，热固性塑料发生聚合，即化学变化，经过硬化定型，变成既不熔融也不溶解的状态。热固性塑料制件成型后，受热不再熔融和软化。热塑性塑料受热软化、熔融，冷却后就固化定型。如果再受热又可软化熔融和冷却固化定型，它们可以反复地加热和冷却，却能保持化学特性不发生变化。

填料。塑料中加入的一些其他材料被称为填料。填料可用来增加体积、改善塑料的性能。塑料中加入填料后，塑料的收缩量会减少，能够很快固化定型而形成最终尺寸。常用木粉作填料。由棉绒纤维制成的棉绒填料可增加塑料的力学性能。将截面面积为 $0.5in^2$（$3.22cm^2$）的棉绒纤维加入塑料中可以增强塑料的强度和抗冲击性能。用石棉纤维作为填料可增强塑料的耐热性和耐火性，而用云母可使塑件具有良好的电绝缘性能。玻璃纤维、硅、纤维素、黏土或坚果壳粉同样可作为填料使用。坚果壳粉常常用于替代木屑粉，以得到更好的表面精度。用短纤维填料能降低塑料成本；而使用长纤维可使塑料性能增强，但价格较贵。除了填料外，还可加入染料、颜料、润滑剂、促进剂和增塑剂等其他材料。加入增塑剂可增加塑料的柔韧性，改善其流动性。通常在塑料原料充模成型前加入填料和改性剂等材料。

塑料的性能。化学家和化学工程师已经发现和研制了很多种类的塑料。在当今的有机化学方面，人造树脂的研究是最为突出的领域之一。塑料的力学性能与金属相比较低，其拉伸强度通常为 $5000 \sim 15000 lb/in^2$（$34.5 \sim 103.4MPa$）。

一些塑料如酚醛树脂，含有多种不同的成分，其他大量有特殊性能的塑料也已被应用。然而，以上只是对塑料进行了简单的介绍。

第 18 课　塑料的压缩模塑

在压缩模塑过程中，塑料原料以粉料或锭料的形式放置在加热的金属模具型腔中。因为分型面处于水平面，上模垂直下行。模具闭合后，预成型加压加热作用一段时间。在压力 $2 \sim 3tf/in^2$（$30.4 \sim 45.6MPa$）和温度 $350°F$（$177℃$）的作用下，使塑料呈半液体状态，充满模具型腔。虽然近年来开发的聚酯塑料可在 25s 内固化，但通常塑料的固化需要 $1 \sim 15min$。最后开模取出塑件。如果塑件中含有金属嵌件，应在注入塑料前，将嵌件放入型腔定位柱上或定位孔中。锭料在装入模腔前应预热，以除去气体，增强流动性，便于充填模具和缩短固化时间。介质加热是加热锭料的便利方法。

因为塑料是直接加入模具型腔的，所以压缩模比其他模塑工艺的模具简单，不需要浇口和流道，可以节省原料。因为清理浇口和流道，对热固性塑件来说是完全损失。用于压缩模塑的压力机通常是立式液压机。较大的压力机要求操作者全神贯注地操作。然而，一个操作者可以同时操作几台小型压力机。压力机要合理放置，以便操作者能够方便地从一台到另一台进行操作，操作者要保证当他再次转到压力机前时，恰好可以开模。

热固性塑料在加热和加压作用下硬化，这个特性决定了其适合压缩模塑和压注模塑。因为这两类模具要交替加热和冷却，所以热塑性塑料不适合用这两种方式成型。为了使热塑性塑料制件硬化及将其从型腔中顶出，就需要将塑件冷却。

压缩模的类型。压缩模可分为四种基本类型，分别是不溢式压缩模、挡环不溢式压缩模、溢式压缩模和半溢式压缩模。在不溢式压缩模中，柱塞式凸模随上模进入下模型腔。因为下模没有挡环或限位装置，柱塞式凸模继续下行以全部压力施加在塑料上，这样就生产出具有良好电性能和物理性能的致密塑件。加入模腔的塑料量须精确计量，因为其影响零件的壁厚。挡环不溢式压缩模与不溢式压缩模类似，只是在预定点处增加了挡环，使柱塞式凸模在预定点停止运动。在这种模具中挡环承受了原应作用在零件上的部分压力。这种压缩模可精确控制零件的壁厚，但是塑件密度的变化是相当大的。在溢式压缩模中，溢料会增加上、下模的合模尺寸。当上模对塑料施压时，多余的熔融物料会从分型面处溢出。对溢料进一步施压，它就会硬化并最后阻挡上模的向下运动。生产致密塑件通常需要填充比计算量稍多的原料来保证足够的压力。这种类型的模具之所以在生产中被广泛采用，是因为它相对比较容易制造，并能够在极限范围内控制塑件的厚度和密度。半溢式压缩模是溢式压缩模和不溢式压缩模的组合类型，除了飞边外，采用的挡环可限制上模的移动。

第19课　传递模塑和注塑（注射成型）

传递模塑。在压缩模塑中，预成型料锭受压而成为流体并充满型腔，这就可能导致细长型芯折断和嵌件移位。同样，对于薄壁的复杂件，难以得到适当的熔融流动状态。为了克服这些困难，开发出了传递模塑。在传递模塑中，塑料在外力作用下以流体状态进入型腔，顺利流经嵌件和细长型芯，进而充满模具型腔（见图19-1）。

用于传递模塑的压注模有两种基本类型，常规流道型压注模和正柱塞型压注模。在流道型压注模中，预成型料锭被放置在型腔上方的一个独立加料室中。一条或多条流道通向模具的分型面，熔料从分型面处通过浇口流向单个或多个型腔。具有浮动中间压板的专用压力机适用于双分型面模具。柱塞直接作用在塑料原料上，使其通过流道和浇口进入模具型腔。温度和压力必须保持一定时间以利于塑料的固化。当塑件固化后，压力卸除，塑件从浇口处与流道分离。柱塞末端鸠尾式的锥形凸出部分带动流道凝料上移。在水平推力作用下，流道凝料就很容易从鸠尾式的锥形凸出部分上取下。在正柱塞型压注模中，由于除去了流道，加料室延伸到模具分型面，并与浇口直接相连。正柱塞型压注模比较常用，因为这种模具的结构不复杂，也不浪费材料。与压缩模塑相比，传递模塑生产出的塑件不仅具有较高的强度、比较均匀的密度以及较为精确的尺寸，而且分型面处需要清理的毛刺也少。

注塑。尽管通过改进，注塑可以用于成型某些热固性塑料，但注塑还是主要用于成型热塑性塑料制件。对于热固性塑料来说，最难解决的问题是将熔融的塑料从料筒注入模具型腔，因为热固性塑料熔体仅在几分钟内就会固化和硬化。注塑的工作原理与压铸法相似，塑料粉粒被加入料斗，在柱塞推力的作用下一定量的塑料进入料筒的加热室。在加热室中，塑料粉粒在加热和加压作用下，变成流体。加热温度通常在 $265 \sim 500℉$ （$130 \sim 260℃$）之间。模具闭合后，柱塞前移，在 $12\,000 \sim 30\,000 \text{lb/in}^2$ （$82.74 \sim 206.84 \text{MPa}$）压力作用下塑料熔体被推入模具型腔。通过循环水对模具进行冷却。当柱塞后退，模具打开时，硬化定型的塑件被推出。注射机可设置为手动操作、自动单循环操作和全自动操作。一般情况下，典型机器每分钟注射4次，塑件质量可达22盎司（0.62kg），在一些机器上，可获得每分钟注射6次的速率。除了表面镀铬外，注射模与压铸模相似。注塑的优点是：

1. 生产效高，适于大批量生产。
2. 可成型具有各种性能的热塑性塑料。
3. 能够成型螺纹、凹槽、侧孔和大型的薄壁件。

第7单元

第20课　机械切削加工基本原理

机械加工是从工件上以切屑形式去掉不需要的材料的工艺方法。如果工件是金属件，这种方法常被称为金属切削或金属切除。美国工业界每年耗资600亿美元用于完成金属切削加工，这是因为大量的产品在制造阶段需要进行切削加工。生产加工范围从相对粗糙或没有精度要求的产品，如铸件或锻件的去黑皮加工，到公差为0.0001in（0.002mm）甚至更小的高精度和高表面质量的产品。可以肯定地说，金属切削加工是最重要的机械加工基本方法。

在过去的80年里，人们通过对金属切削工艺的大量研究和实验，增进了对加工工艺性和被加工表面特性的了解。尽管这种努力已经使机械加工效率得到提高，但由于机械加工工艺的复杂性致使在获取金属切屑成形的完整理论方面进展缓慢。

是什么原因使得金属切削工艺研究如此不寻常和困难呢？

不同的材料表现出不同的特性。

加工过程不均衡，除刀具限制外不受其他约束。

应变程度非常大，应变速率很高。

加工过程对刀具几何形状、材料、温度、环境（切削液）和加工动力学特性（颤动和振动）等均非常敏感。

这一章的目的就是如实地将这方面展示给生产实践中的工程师。

基本的金属切削加工方法主要有7种：刨削、车削、铣削、钻削、锯削、拉削和磨削（研磨加工）。对于所有金属切削加工，都必须正确区分切削速度、进给量和背吃刀量之间的关系。现以车削加工为例介绍这些术语（见图20-1）。通常，切削速度（V）是指切削时的主运动速度，在车削加工中是旋转的工件相对于固定不动的刀具的速度。通常以英尺每分钟或英寸每分钟（in/min），或米每分钟（m/min），或者米每秒（m/s）来表示。切削速度（V）在图20-1中用加粗的箭头表示。进给量（f_r）是工件每转一周或一个行程中刀具切除材料的量。在车削加工中，进给量以in/rad表示，刀具沿平行于工件旋转轴线方向进给。根据不同加工方法，进给量单位为in/r、in/min或in/刃。图20-1中以长箭头表示进给。在车削加工中，背吃刀量（DOC）是第三个尺寸，以字母d表示。在车削中，背吃刀量是刀具切入工件表面的距离，是工件初始直径D_1和加工后直径D_2差值的一半：

$$d = \frac{D_1 - D_2}{2} \tag{20-1}$$

切削速度是切削刃在工件表面的圆周速度，以ft/min表示。车削中每一转的进给量是刀具在平行于工件旋转轴方向上运动的长度。背吃刀量是第三个尺寸。L为切削长度。

旋转工件的表面速度与工件外形尺寸有关：

$$V = \frac{\pi D_1 N}{12} \tag{20-2}$$

式中 D_1——工件原始直径（in）；

V——工件表面速度（ft/min）；

N——工件每分钟旋转次数（r/min）。

另一个被编者省略的图表示用于车削加工工艺的典型机床——车床。工件被固定在夹紧装置上，在这个例子中用三爪卡盘夹紧工件并相对刀具做旋转运动。卡盘安装在车床主轴上，主轴通过电动机带动齿轮驱动。切削刀具用于直接加工工件，是加工过程中最关键的因素。在确定切削速度和进给量之前必须选择刀具材料和刀具几何形状。机床、刀具和夹紧装置通常在不同的公司制造，刀具成本与机床相当或高于机床成本。

第 21 课　在车床上车削和镗削的基本原理

车削是对圆柱体和圆锥体外表面的机械加工方法，通常在车床上完成。在这一章的后面将给出车床的照片和图解。图 21-1 所示为车削一个简单圆柱体表面时工件与刀具之间的运动关系。工件旋转，单头车刀纵向进给。如果刀具与旋转轴成一定角度进给，则形成外圆锥形表面，这就叫作锥体车削。如果刀具与旋转轴成 90°方向进给，所用的刀具宽度比车削的宽度宽，就形成了一个平面，这种操作叫作端面车削。

用具有特殊形状刃口的刀具沿工件径向或轴向进给，则可以在有限长度内车削出外圆柱面、锥面、不规则表面。工件的表面形状由车刀的形状和尺寸所确定。这种机械加工被叫作成形车削。如果车刀朝工件轴线切下去，工件则一分为二，这种操作叫切断，通常使用形状简单而薄的车刀，在轴颈车削或部分切掉车削中也用类似的车刀。

镗削是由车削演变而来的，实质上镗削是内部车削。镗削加工可采用单头镗刀加工出内圆柱体或圆锥体。但是镗削不能加工出孔，而是将已有孔加工到特殊尺寸。镗削能够在大多数可以旋转的机床上进行，而且还可以让刀具旋转而工件保持不动。而且还有专用机床可以镗、钻、和铰削，但不能车削。在车床上还可以进行螺纹加工和滚花等操作。此外，钻孔、铰孔和车圆锥也能够在工件的旋转轴上完成。

车床的主要工作是车削。车刀自右向左进给时将产生切削力。切削力应指向主轴箱，以迫使工件紧靠在卡盘上，这种方式有利于工件的夹持。

如果工件要求非常光洁的表面和精确的尺寸，通常先安排一次或多次粗车之后再安排一次或多次精车。粗车可以采用适当的切屑厚度、刀具寿命、车床动力和工件允许的其他参数。在相反的工艺过程中，大的背吃刀量和小的进给量更具优势，因为少切几刀减少了溜板往返的时间和重磨刀具的时间。

当工件表面坚硬时，如有氧化铁皮的铸件或热轧材料，初次粗车一定要有足够的深度以穿破表面硬化材料。否则整个切削刃全程工作在硬质研磨材料中，车刀将很快变钝。如果材料表面非常硬，在第一次粗车时要相应地降低切削速度。

精车要轻，背吃刀量通常小于 0.015in（0.38mm），进给要尽量精确，以达到表面粗糙度要求。有时采用特殊的精车刀具，但通常情况下粗车和精车使用相同的车刀。多数情况下至少要求一次精车。然而如果有特殊的精度要求，可进行两次精车。如果由人工控制直径，则应在完成全部精车之前，先车一小段，如 1/4 in（6.35mm）长，检查一下直径。这是因为先前是在粗糙的表面用千分尺测量尺寸，重置测量工具在光滑的表面上精确地测量是必要的。

车削中的主运动是工件的旋转和刀具沿平行于旋转轴线的进给运动（见图21-1）。工件的转速N（r/min）由切削速度V确定。进给量f_r以in/r表示。背吃刀量d用下式计算：

$$d = \frac{D_1 - D_2}{2} \tag{21-1}$$

切削长度L是刀具平行于旋转轴移动的距离，再加上车刀切入余量或超越行程A，A是允许刀具进入或退出的距离。

一旦基于给定的工件材料、刀具材料而选定了切削速度、进给量和背吃刀量，那么机床的转速可以由下式确定：

$$N = \frac{12\,V}{\pi\,D_1} \tag{21-2}$$

（适用于较大直径，）式中的数值12将单位英尺转换为英寸。

镗削总是扩大已经存在的孔，这个孔可能是钻出来的，也可能是铸造时由型芯形成的。同样重要并且存在的问题是镗孔可以使孔的中心与工件的回转中心统一，以纠正由于钻孔使中心线漂移带来的偏差。同轴度是镗孔的一个重要特性。

当在车床上镗孔时，通常工件装在卡盘上或花盘上，可以镗直孔、锥孔或不规则孔。图21-2所示为镗孔刀具和工件之间的关系。镗削基本上是内表面切削，而刀具沿平行于工件的旋转轴线进给。

第22课　铣　　削

铣削是以切屑方式去除金属材料以获得加工面的一种最基本的加工方法。铣削加工时，工件向旋转的刀具送进。有时工件保持静止，而刀具进给。几乎在所有的情况下，铣削使用多切削刃刀具，因而具有较高的材料切削率。通常刀具或工件一次行程就能获得所希望的表面，而且表面质量较好，因此铣削特别适合并广泛用于大批量生产。生产中应用多种铣床，其范围包括生产车间用于常规加工的简易铣床和万能铣床，也包括在工模具车间使用的用于大批量生产的各种高度专业化铣床。毫无疑问，更多的平面加工将采用铣削而不用其他的机械加工方法。

铣削加工的刀具叫作铣刀。在刀具周边等距离分隔的刃口断续切入并加工工件，这叫作不连续切削。

铣削加工可分为两类：圆周铣削和端面铣削。每一类还有多种形式。圆周铣削时，加工表面依靠分布在铣刀体周边的切削刃切削成形，如图22-1所示。待加工面平行于刀具旋转轴。用这种方法可以进行平面铣削和成形表面铣削，成形表面的横截面形状与铣刀的轴向轮廓一致。这种加工常被称为面铣，常在卧式铣床上完成。铣削平面时，铣刀以转速N（r/min）旋转，工件在工作台上以速度f_m（in/min）进给。

与其他加工过程类似，切削速度V和每齿进给量由工程师和铣床操作者选择。如前所述，这些变量取决于工件材料、刀具材料和特殊加工要求。切削速度指铣刀齿切削刃上的速度。铣床主轴的转速由工件表面的切削速度确定，当铣刀直径的单位为英寸时，计算公式如下：

$$N = \frac{12V}{\pi D} \tag{22-1}$$

背吃刀量t，简单定义为原表面与铣削后表面之间的距离，如图22-1所示。切削宽度是铣刀宽度或工件宽度，以英寸计量，用W表示。切削长度L是工件长度加上为刀具接近工件和超越工件的修正量L_A。工作台的进给量与铣刀每一齿在铣刀每一转中切削的金属量（即每齿进给量）有关，进给量以f_m表示，单位为in/min，而每齿进给量以f_1表示，其关系为

$$f_m = f_1 Nn \tag{22-2}$$

这里n是齿状铣刀上的齿数（齿/转）。

在端面铣削中，加工面与铣刀轴垂直（见图22-2），大部分切削由铣刀齿周边的切削刃完成，铣刀齿端面的切削刃可完成精加工。端面铣削既可以在卧式铣床上完成，也可以在立式铣床上进行。

第8单元

第23课 电解加工

电解加工（ECM）的基本工作原理是：工件为阳极带正电荷，工具为阴极带负电荷，在电解液中工件与工具之间发生电荷与金属材料交换。在这样的条件下，阳极被分解，而此时工具电极则不受影响。金属去除量可以遵照法拉第定律计算

$$V = CIt$$

式中 C——根据工件材料而定的系数；

I——工具与材料间的电流；

t——电腐蚀时间。

电流强度取决于工具电极与工件之间的放电间隙、电腐蚀面积、电解液的导电性，以及供电电压。电极与工件之间保持的工作间隙，既避免了物理接触，又能确保发生电解。电解液（如NaCl或$NaNO_3$等的水溶液）用泵送入工作间隙，同时由于间隙处高密度的能量电解液也必须作为冷却液使用。加工中从工件上腐蚀的金属碎屑须用滤网或离心机使其与电解液分离。

电解刻模机。图23-1为电解刻模机主要组成部分示意图。进给装置依据去除金属的速率要求推动工具电极向工件运动。当加工内表面时，工具电极的形状和尺寸将成为设计的主要问题。间隙不是常数，它是以腐蚀表面的状态和工具的推进速度为变量的函数。例如，如果要刻制一个圆柱形孔，一个简单的圆柱形工具电极（见图23-2）就不适合了，这将导致间隙尺寸持续增加而电流强度成比例降低（图23-2左图）。用一个表面适当绝缘的电极（图23-2右图）将遏制损伤圆柱侧面的过度腐蚀。

电解机床各元件的功能总结详见表格。由于在电解加工中采取小尺寸间隙，所以必须使用高压力（大于20bar，$1bar = 10^5 Pa$）的电解液才能有适当的流速，从而有效地冷却和带走金属腐蚀物。在$10000 mm^2$的腐蚀面积中，要作用超过20kN的力，机床的工具电极进给系统和机床结构都能够承受如此的压力。由于电解液是由腐蚀性盐溶液构成，机床所有可能接触到电解液的部件都必须有防腐性。有一个重要的辅助装置是短路断电装置，当发生工具电极与工件的间隙不充足，以及腐蚀金属碎屑清除不当时，则立即停止继续供电。

加工中小尺寸工件的电解刻模机采用开式的"C"型结构，而用于加工大型工件的机床采用闭式的"O"形结构。

工具电极进给方向首选垂直方向。图23-3所示电解机床采用拼装系统，机架和工件夹持头可以实现与床身的各种结合。大多数情况下工具电极进给采用电液驱动，也可以由电装置实现，例如把伺服电动机连接到一个带螺母的滚珠丝杠上。

第24课　电火花加工

电火花加工过程中，工具电极与工件之间发生放电而使工件材料腐蚀。由于放电时间短，但温度高，放电点的金属微粒熔化，部分汽化，并通过机械和电磁力被去除。工作介质是非导电液体，它能清洗掉电腐蚀材料，同时也用作冷却剂。

如同ECM加工，EDM是一种复制加工，且工具与工件之间没有物理接触。然而与ECM相反，在EDM加工中会对工具有一些腐蚀，因此在工具设计中要充分考虑这种腐蚀，以保证加工精度。

电火花加工与电解加工之间的另一个区别表现在电火花工具的加工进给量不固定，加工间隙的大小是依据去除金属的速率与该间隙的加工状态来确定的。

电火花刻模机。电火花刻模机的结构原理如图24-1所示。电火花腐蚀发生在装夹有工件的充满绝缘介质的容器内。电极的进给控制是通过电液装置或机电伺服系统完成的。电腐蚀电能由电腐蚀发电机提供。过滤器从工作介质中分离出电腐蚀材料。在图24-1的左上部分，以放大图解形式说明一个单独的电火花放电。施加的电压在放电的初期使间隙电离。在放电区域内强度最高点，形成放电通道，放电电流通过通道流动。每个放电通道的端部金属熔化，在它的周围气泡膨胀。当放电电压完全卸荷时，通道崩溃，熔化的金属材料汽化，仿佛发生微小爆炸。放电的结果是在电腐蚀工件表面形成了不规则的凹坑和伤疤。

电火花线切割机床。电火花腐蚀加工的一个重要应用是用线电极切割金属。这种工艺用于制造有孔隙的切割刀具和为EDM制造工具电极。图24-2为加工原理图。切割工具是细的黄铜或青铜丝，在切割加工时它伸入工件但却没有与工件接触。因为线电极要受电火花腐蚀，因此要持续不断地更换新电极丝。从图24-3可以看到电极丝进给装置。依据加工工件的要求和工件尺寸来调节金属丝的张紧程度、损耗率和金属丝支撑臂达到的位置。电火花线切割加工常用的工作介质为去离子水，用冷却喷头喷向工作区域。

根据工件轮廓的加工要求，装夹工件的工作台和控制线电极进给的滑块的位置必须准确确定。线电极对工件的进给没有恒定的速度，但在整个加工过程中必须遵照放电间隙存在条件而变化，而这些条件取决于线切割工艺及电火花线切割机床。

第9单元

第25课　数控机床分类

数控机床通常可以按如下方法分类：

1. 按动力驱动方式：液压驱动、气动和电力驱动。

2. 按加工刀具控制方式：点位控制和连续轨迹控制。

3. 按定位系统：增量定位和绝对定位。

4．按控制环：开环和闭环。

5．按坐标定义方式：右手坐标系和左手坐标系。

动力驱动。一台优质的数控机床最值得注意的性能是它的动力源。动力源通常决定了机床加工的能力，从而决定了各种应用的可行性。三种主要的动力源分别是：液压传动、气动和电动。

大多数大功率机床常用液压驱动。液压能够传递很大的力，这样机床滑块能以更稳定的速度运动。但这个优点的代价是液压传动相对于电力驱动或气压驱动成本较高。另外，液压传动要求附加外围设备，如油箱、阀门等。液压传动的最大缺点是液压设备引起的噪声及漏油引起的污染。

气动机床通常是最便宜的替代方式。车间里的压缩空气压力达 $90 \sim 100lb/in^2$（$0.62 \sim 0.69MPa$），可以输送给机床驱动机床运动，这是气动机床的另一个优点。正常情况下气动机床的每一个轴仅在其终点受控。然而由于计时和时序的变化，可能使编程设置变化无穷。气动驱动的特点是机床运动通常不平稳，典型现象是在高速加速或减速时发出"砰砰"的响声。

电气驱动最适于精密工作要求或较精密控制要求。电气驱动数控机床具有精确的运动控制功能。电气驱动装置有两种主要类型。一种是步进电动机。步进电动机能够在控制器输出的每一个电压脉冲控制下精确转动一个角度。步进电动机的运动非常精确，只要提供的转矩负载不超过电动机的设计极限。

鉴于步进电动机固有的精确性，其常用于开环系统中。另外一种电气驱动是伺服电动机驱动。这种电动机作为驱动元件总是与进给反馈环结合，将驱动元件信息从驱动器反馈到控制器。不间断地监视运动位置，并通过提供与位置变化相适应的电压或电流信号迅速纠正偏差，直到位置和速度偏差降为零。伺服驱动装置是连续位置控制装置，这些位置由编码转换器、感应同步器、分解器或其他类似的反馈装置测量。伺服电动机可提供平稳及连续的可控制运动。

运动控制。点位控制（PTP）系统利用坐标系控制刀具从一点移动到另一点。每个刀具的坐标轴可分别进行控制，轴系的运动可以是在一个时间一个轴运动（见图 25-1a），也可以是每个轴以恒定的速度运动的多轴运动（见图 25-1b）。运动的控制总是由程序设定的点定义，而不是由点之间的路径确定。这种控制系统最简单的应用实例是数控钻床。在钻床上，工件或刀具沿两个轴运动，直到刀具的中心到达要求的孔的中心位置。假设钻头在从一点到下一点的运动过程中，其路径和进给量均不重要，这样从路径的起始点到终点就不用控制。位置数据由坐标值给出。快速移动通常是点位控制操作的特点，即使在连续控制系统中也是如此。

第 26 课　加工中心

加工中心的类型。有两种主要类型的加工中心：水平主轴型加工中心和垂直主轴型加工中心。

水平主轴型（卧式加工中心）。

1．移动立柱式加工中心通常配备两个可以安装工件的工作台。使用这种类型加工中心，当一个工作台用于加工工件时，操作者可以在另一个工作台上安装工件。

2. 固定立柱式加工中心有一个托盘传送装置。托盘是可以移动的工作台，在上面可以安装工件。当工件加工完成以后，工件和托盘被转移到传送装置，然后传送装置运转，带动一个新的托盘和工件进入加工位置。

垂直主轴型（立式加工中心）。立式加工中心采用带滑动导轨的床鞍结构，利用立式床头滑动替代了主轴套的运动。

CNC 加工中心的组成部分。CNC 加工中心主要由床身、床鞍、立柱、工作台、伺服电动机、滚珠丝杠、主轴、刀库和机床控制单元组成。

床身。床身通常用高质量铸铁制成，以使机床具有高刚度，同时在繁重的加工任务下仍保持高的精度。床身导轨表面经过硬化并研磨，对全部直线轴提供刚性支撑。

床鞍。床鞍安装在硬化和研磨过的床身导轨上，为加工中心提供 x 轴方向直线运动。

立柱。立柱安装在床鞍上，它设计成具有高的抗扭强度以防止在加工过程中扭转或挠曲变形，立柱为加工中心提供 y 轴方向直线运动。

工作台。工作台安装在床身上，为加工中心提供 z 轴方向直线运动。

伺服系统。伺服系统由伺服电动机、滚珠丝杠、定位反馈编码器组成。伺服系统为 x、y、z 轴滑块提供快速和准确的运动与定位。反馈编码器安装在滚珠丝杠的端部，形成闭环系统，它能保持单向 + 0.0001in（0.003mm）的高位置重复精度。

主轴。主轴的转速范围为 $20 \sim 6000 \mathrm{r/min}$，由程序控制 $1 \mathrm{r/min}$ 的增加量。主轴可以采用固定位置（水平方向），也可以倾斜或偏转，以提供附加轴。

换刀装置。有两种基本类型的换刀装置：垂直换刀装置和水平换刀装置。换刀装置能够预先储存一定数量的刀具，这些刀具被零件的加工程序自动调用。换刀装置中刀具的更换通常是双向的，从而允许以最短的距离任意地存取刀具。自动更换刀具的时间通常仅为 $3 \sim 5 \mathrm{s}$。

MCU。操作者可以通过 MCU（机床控制单元）完成各种操作，如编程、加工、诊断、刀具和机床状态监控等。MCU 根据加工者的技术要求变化；新型 MCU 变得更精确，使机床更可靠，使加工更少地依赖于操作者的技能。

机床坐标轴。加工中心可能对 NC 机床造成最大冲击，因为工件仅需一次装夹就能够完成工件所有侧面的各种加工。当完成一系列操作和程序时，五轴加工中心显示可以应用的轴是：

x 轴　　　　直线运动

y 轴　　　　直线运动

z 轴　　　　直线运动

A 轴　　　　倾斜或偏转主轴

B 轴　　　　转动工作台

第 10 单元

第 27 课　计算机在模具设计中的应用

术语 CAD 既可指计算机辅助设计又可指计算机辅助绘图。实际上这两种含义的 CAD，模具设计师都会用到。

CAD 计算机辅助设计意味着用计算机及其外设装置简化设计过程和提高设计工艺性，而 CAD 计算机辅助绘图则指用计算机和外围设备制作设计过程文件和图样。设计文件通常包括初步设计图样、工件图样、零件列表和设计计算。

一个 CAD 系统，无论是指计算机辅助设计系统还是计算机辅助绘图系统，都由三个基本部分组成：（1）硬件；（2）软件；（3）用户。一台典型的 CAD 系统硬件组成包括处理器、显示器、键盘、数字化处理器和绘图机。CAD 软件由能够完成设计和绘图功能的程序组成。用户是模具设计师，他用硬件和软件简化设计过程，提高设计工艺。

直至 20 世纪 80 年代，CAD 在工业范围内都没有实质性地广泛出现。现在 CAD 这个概念已被我们所熟悉。尽管在过去几十年中 CAD 有非常大的改变，但最早出现 CAD 是 20 世纪 50 年代中期。那时个别的第一代计算机才包含有图形显示系统，而现在图形显示系统是每一个 CAD 系统必须拥有的部分。

图形显示系统代表了工具设计领域朝着与计算机结合走出的真实的第一步。绘图机在图形表达上实现了第二步。随着 20 世纪 60 年代早期出现的数字化输入板，我们现在熟知的 CAD 硬件开始成形。紧接着计算机硬件的发展计算机图形处理软件也快速发展。

早期的 CAD 系统又大又笨重，且价格昂贵。事实上由于如此昂贵，仅少数大公司能够承受。在 20 世纪 50 年代末及 60 年代初期，CAD 看起来有趣、新颖，却不切实际，在模具设计领域仅有有限的潜在应用可能性。然而随着 20 世纪 70 年代硅片的引进，计算机开始在模具设计领域占据自己的位置。

由于计算机所需的集成电路均制作在小小的硅片上，因此组装起来的全功能计算机与电视机一样大。"小型"计算机具有全功能计算机的全部性能，但体积要小得多，且相当便宜。甚至更小的微型计算机很快也将如此。

20 世纪 70 年代，CAD 硬件和软件技术继续进步。更大的进步则开始于 20 世纪 80 年代，CAD 的生产和市场化使其成为发展飞快的行业。而且今天 CAD 已经从不切实际的新玩意转变成为最重要发明中的新成员。到了 20 世纪 80 年代，众多 CAD 系统可以应用于从微型计算机系统到小型计算机和主机系统。

第 28 课　CAD/CAM

计算机辅助设计/计算机辅助制造（CAD/CAM）指的是计算机系统整体融入产品设计和制造过程，以提高生产效率。CAD/CAM 系统的核心是设计终端和与之相关的硬件，如计算机、打印机、绘图机、纸带穿孔机、纸带阅读器及数字转换器等。设计过程可一直在终端监控，直至完成。如需要可以进行硬拷贝。在产品进行加工、实验和质量控制时，计算机磁带或其他含有设计数据的媒介物引导计算机控制加工机床。

CAD/CAM 软件是一批储存在系统中的计算机程序，它们控制系统的各种硬件完成特定任务。软件的一些实例如创建 NC 机床路径、调集材料单或在有限元模型上创建结点和元素等工作的程序。一些软件包被称为软件模型，可以分为四类：（1）操作系统，（2）通用程序，（3）应用程序，（4）用户程序。尽管还有其他种类的软件，但这四种分类足以解释在开发 CAD/CAM 软件系统中错综复杂的情况。

操作系统是为特殊计算机或分级别计算机编写的程序。为方便和有效地操作计算机，程序和数据存储在计算机内存中。操作系统特别与输入/输出（I/O）设备有关，如显示器、打印机和纸带穿孔机。大多数情况下操作系统随计算机一起供给。

尽管有人认为没有全能的通用软件，但是有些软件比其他一些软件要通用。例如用像FORTRAN这样的高级语言编写的绘图程序可以创建像线条、圆和抛物线这样的几何实体，并可把它们组合起来进行设计。设计范围可从印制线路到设计钻模和夹具。

应用程序是为专门目的或特殊用途而开发的。第一个专用语言是1956年的自动编程工具（Automatically Programmed Tools，缩写APT）。APT用于简化开发输入NC机床的NC程序，如图28-1所示。与CAD/CAM相关的应用程序的其他例子是为一些专门用途开发的程序，如用于创建有限元网格和平面模型开发或"无挠曲"的金属板零件的软件。这些软件通常与系统一起购买或由软件供应商提供。

为了特殊的输出要求，CAD/CAM中的用户程序要求高度专业化的打包。例如，当用户输入如齿数、节圆直径等参数后，用户程序可以自动地设计齿轮。另一个程序可以根据给出的刀具参数、材料、背吃刀量等数据计算出最佳进给和切削速度。这类程序通常由用户利用软件供应商提供的软件模型或利用通用软件开发出来。不是所有的CAD/CAM软件包都有这样的用户程序，尽管应用它们可以达到相当节省的目的。

计算机绘图。计算机绘图系统积累并储存了与精定位、尺寸描述文本和每一个设计元素的特征等相一致的有关实际数据。这些与设计有关的数据帮助用户完成复杂的工程分析，建立材料清单，生成分析报告，在零件被送去加工前检测出不合理的设计问题。

用计算机二维绘图能够建立三维线框模型和实体模型。

线框模型。简单的线条结构曲线图是最便宜的表现几何模型的形式。线条图能有效地表达平面图形特征和保持模型连贯。然而，当开发复杂的产品模型时，线条图就不能很好表达，而实体模型则可以解决线框模型中的大部分问题。

实体模型。有三项基本技术用于创建实体模型：建立实体几何体（CSG）、边界表达（B-Rep）和分析实体模型。

在CSG方式中，各种典型的几何体，如圆柱、球体、锥体，由布尔代数结合在一起进行创造性地设计。

在B-Rep方法中，先定义零件轮廓，然后扫描，扫描既可以是线性的，也可以是放射性的，封闭的面积都可以描述为实体形式。

分析方法。这种方法与B-Rep方法类似，但是它在设计图形过程中增强了有限元模型的创造性。商业化的软件包并不严格地用这种方法或另一种方法。例如，CSG软件包可能用B-Rep技术创建最初的几何模型，而B-Rep软件包或分析软件包可能用布尔代数去除某些几何模型，如在设计的几何体中用去除圆柱体或圆锥体而形成一个孔。

计算机辅助制造（CAM）的核心围绕着四个领域：数字控制、制订工艺过程、机器人和工厂管理。

数字控制。在CAM领域中，NC的重要性是计算机能够直接从几何模型或零件创建NC程序。目前这种自动创建能力还局限于高度对称的几何体和其他一些特殊零件。然而，在不远的将来，一些公司将完全不用画图，而是通过数据库直接将设计的零件信息输送到制造场地。由于不再使用图样，从而因使用图样所带来的许多问题也会随之消失。这是因为此时设计和制造所使用的计算机模型都是由一个通用集成数据库形成的。这样做甚至不用在乎设计部门和制造部门距离非常遥远，因为在本质上再远也远不过放置在工作台上的各个终端机间的距离。

制订工艺过程。制订工艺过程包括从生产开始到完成各个生产程序的详细描述。一个与CAM 相适应的制订工艺过程系统能够不用人辅助直接从几何模型数据库生成工艺过程文件。

机器人。CAM 的许多优点集中在机器人进入制造系统中，如在装配线、焊接生产线和喷漆线上应用机器人。

工厂管理。工厂管理使交互式工厂数据采集系统及时从生产场地得到信息；同时，利用这些数据计算出制造的先后顺序、动态地确定下一步要做的工作，从而确保正常地执行标准制造程序。这个系统还能够无须请示计算机程序专家直接修改工艺，以满足特殊需要。

第 29 课　快速原型制造

原型是指"与复制、仿造、表现、新试样或改型有关的最初的物体"（摘自韦氏字典）。快速原型是"一种针对直接来自于数字作图（典型的 CAD 模型）的任意形状的三维尺寸的实体零件，采用快速、高效自动化以及完全柔性工艺制造的制造方法"，见 1992 年 10 月《快速原型报道》。

近年来以及在写这篇文章的过程中，多种类型的快速原型制造已经出现。这方面的技术发展包括光固化成形（SL），选域激光烧结成形（SLS），熔融沉积成形（FDM），分层实体制造（LOM），以及三维打印（3DP）。它们都能够借助计算机辅助设计（CAD）数据库生成物体。

这些技术的应用，使产品的研制时间缩短，同时还使制造各种类型、尺寸产品的柔性得以改进。

快速原型制造的特征。快速原型制造工艺由两步组成：第一步，借助立体几何模拟装置模拟零件，然后在计算机上将其切成一系列平行的二维截面层；第二步，将二维截面层数据直接用于指示机器从下向上逐层制造零件。快速原型制造的一个共有的重要特征是通过分配材料而不是去除材料来生产原型零件。

运用这一技术改进产品研制有以下三个方面的好处：

（1）在工程设计方面

设计者在很短时间内，运用 CAD 设计技术设计现实中复杂产品的可视模型，即产品的原型，从而使工程师们能够很快地对设计出的产品进行评价。

没有实质上的工具和劳动成本，快速原型制造能快速制造出原型，并且还能在限定的时间范围内用可能提供的成本改进产品的质量。

（2）在制造方面

在设计阶段就能提供一个有形产品，这样可以加速工艺和工具的设计，尽可能减少设计图中存在的问题。

（3）在市场营销方面

通过对原型概念、设计思想以及公司的生产能力的说明，我们能够得到用户的实际反馈意见，有利于及时地改进产品的设计和促进产品销售。

光固化成形（SL）通过激光固化液态光敏树脂实现快速原型，1987 年由美国 3D systems 公司商业化推出。

光固化成形装置是采用激光束，在液态树脂光聚合物溶池表面通过扫描截面层产生原型的。如图 29-1 所示，激光作为一个点光源扫过液态树脂表面，使底层首先固化。固化层随升降台下移约 $50 \sim 375 \mu m$（$0.002 \sim 0.015$ in，in $= 25.4$ mm），这取决于所需要的精度。随后下层光敏树脂被光固化，并且熔融到其下一层。

可光固化液态树脂经研制可用于印刷和家具油漆/密封剂。由激光提供直接能源。使用这种能源，可使最初的乙烯基单体（小分子）聚合成大分子，且强度提高。

为了形成一薄片单层，激光束扫描出第一层的边界轮廓轨迹，这与计算机数控机床加工中的仿形或 Z 字形刀具运动不同。仿佛一个大的弹性带或环位于表面，这称为设立边界。其次，阴影线区域被填满，引起最后的胶化和凝固。每一层形成后，顶面扫描移向下一层，在前一层的上面形成新的一层。用这种方法，从底到顶一层一层建立起模型。然而，对于千分之几英寸的精度要求，必须要对工艺进行仔细的设计。

当所有的层完成后，原型的 95% 固化了。辅助固化（后固化）可以使原型完全固化。辅助固化在荧光烘箱中完成，在这里大量紫外线流射向被固化体（原型）。

第 30 课　先进制造技术

柔性制造。在现代制造装置中，柔性是一个重要的特征。这意味着制造系统具有多样性和可适应性，同时具有处理相对大批量的流水线生产的能力。柔性制造系统的多样性表现在它可以生产各种各样的零件。由于它能够快速地改进，以生产完全不同的零件，所以非常适用。

柔性制造系统（FMS）是由计算机控制并具有工具运送能力的自动化物料输送系统为之服务的一台机器或一组机器。由于它的工具输送能力和计算机控制能力，系统能够连续地进行改装，以制造各种各样的零件。这也是为什么称之为柔性制造系统的原因。典型的柔性制造系统包括以下几个方面：

(1) 工艺装备，例如机床、装配站及机器人。

(2) 物料运送设备，例如机器人、传送装置和 AGVS（自动引导小车系统）。

(3) 通信系统。

(4) 计算机控制系统。

柔性制造代表向整体集成制造目标迈进的一个重要的阶段。它包括自动化生产工艺的集成。在柔性制造过程中，自动化的制造机器（即车床、铣床、钻床）和自动化物料输送系统同时共享经由计算机网络传送的信息。

借助于集成多个自动化的制造概念，柔性制造向完全集成制造目标迈出了重要一步，这些概念包括：

(1) 单一机床的计算机数字控制。

(2) 制造系统的分布式数字控制。

(3) 自动化物料输送系统。

(4) 成组技术（零件族）。

当把这些自动化工艺、机器及概念集成在一个系统时，就构成了柔性制造系统。在柔性制造系统中，人和计算机起着主要的作用。当然，人的劳动量是远远少于手工操作制造系统中人的劳动量。然而，在柔性制造系统的操作过程中人仍然起着最重要的作用。人的任务包括以下几个方面：

（1）设备故障检修、维护以及修理。

（2）刀具更换和调整。

（3）系统加载和卸载。

（4）数据输入。

（5）零件程序的变换。

（6）程序开发。

柔性制造系统的构成。 一个柔性制造系统由四个主要部分构成：机床、控制系统、物料运送系统、操作人员。

Appendix B   Tables of Weights and Measures

Table B-1   Basic Units in the International Unit System
(国际单位制的基本单位)

Name of Quantity 量的名称	Name of Unit 单位名称	Symbol of Unit 单位符号
length 长度	metre 米	m
mass 质量	kilogramme 千克	kg
time 时间	second 秒	s
electric current 电流	ampere 安［培］	A
thermodynamic temperature 热力学温度	kelvin 开［尔文］	K
amount of a substance 物质的量	mole 摩［尔］	mol

Table B-2   Derived Units with Special Names in the International Unit System
(国际单位制中具有专门名称的导出单位)

Name of Quantity 量的名称	Name of Unit 单位名称	Symbol of Unit 单位符号	Other Form 其他表示
force 力，重力	newton 牛［顿］	N	$kg \cdot m/s^2$
frequency 频率	hertz 赫［兹］	Hz	s^{-1}
pressure, intensity, stress 压力，压强，应力	pascal 帕［斯卡］	Pa	N/m^2
energy, work, heat 能量，功，热	joule 焦［耳］	J	$N \cdot m$
power, radiant flux 功率，辐射通量	watt 瓦［特］	W	J/s
celsius temperature 摄氏温度	degree Celsius 摄氏度	℃	
capacitance 电容	farad 法［拉］	F	C/V
electrical resistance 电阻	ohm 欧［姆］	Ω	V/A
potential, voltage, electromotive 电位，电压，电动势	volt 伏［特］	V	W/A

Table B-3   Non-international Units Stipulated by the State
(国家选定的非国际单位制单位)

Name of Quantity 量的名称	Name of Unit 单位名称	Symbol of Unit 单位符号	Conversion 换算关系
time 时间	minute 分 hour［小］时 day 天（日）	min h d	1 min = 60 s 1 h = 60 min = 3600 s 1 d = 24 h = 86400 s

Name of Quantity 量的名称	Name of Unit 单位名称	Symbol of Unit 单位符号	Conversion 换算关系
plane angle 平面角	second［角］秒 minute［角］分 degree［角］度	(″) (′) (°)	 $1' = 60''$ $1° = 60'$
speed of rotation 旋转速度	rotation per minute 转每分	r/min	$1 \text{r/min} = (1/60) \text{ s}^{-1}$
length 长度	nautical mile 海里	n mile	$1 \text{n mile} = 1852 \text{ m}$（航海）
speed, velocity 速度	knot 节	kn	$1 \text{kn} = 1 \text{n mile/h}$
mass 质量	ton 吨	t	$1 \text{t} = 1000 \text{ kg}$
cubic measure 体积	litre 升	L (l)	$1 \text{L} = 1 \text{ dm}^3 = 10^{-3} \text{ m}^3$
energy 能	electron volt 电子伏	eV	$1 \text{eV} \approx 1.6 \times 10^{-19} \text{ J}$
acreage of land 土地面积	hectare 公顷	hm² (ha)	$1 \text{hm}^2 = 10000 \text{ m}^2$

Table B-4 Common Non-legal Units and Legal Units
法定计量单位与常见非法定计量单位的对照和换算（部分）

Name of Quantity 量的名称	Legal Unit 法定计量单位		Common Non-legal Unit 常见非法定计量单位		Conversion 换算关系
	Name of Unit 单位名称	Symbol 符号	Name of Unit 单位名称	Symbol 符号	
Length 长度	kilometer 千米（公里）	km	mile 英里	mile	$1 \text{mile} = 1760 \text{yd} = 5280 \text{ft} = 1.609 \text{km}$
	meter 米	m	yard 码	yd	$1 \text{yd} = 3 \text{ft} = 0.9144 \text{ m}$
	decimeter 分米	dm	foot 英尺	ft	$1 \text{ft} = 12 \text{in} = 30.48 \text{ cm}$
	centimeter 厘米	cm	inch 英寸	in	$1 \text{in} = 2.54 \text{ cm}$
	millimeter 毫米	mm			
	micrometer 微米	μm			
Area 面积	square kilometer 平方千米（公里）	km²	are 公亩	a	$1 \text{a} = 100 \text{ m}^2$
	square metre 平方米	m²	square mile 平方英里	mile²	$1 \text{ mile}^2 = 2.58999 \times 10^6 \text{ m}^2$
	square centimeter 平方厘米	cm²	acre 英亩	acre	$1 \text{ acre} = 4840 \text{ yd}^2 = 4046.856 \text{ m}^2$
			square yard 平方码	yd²	$1 \text{ yd}^2 = 9 \text{ ft}^2 \approx 0.836 \text{ m}^2$
			square foot 平方英尺	ft²	$1 \text{ ft}^2 = 144 \text{ in}^2 \approx 0.093 \text{ m}^2$
			square inch 平方英寸	in²	$1 \text{ in}^2 = 6.4516 \times 10^{-4} \text{ m}^2$

（续）

Name of Quantity 量的名称	Legal Unit 法定计量单位		Common Non-legal Unit 常见非法定计量单位		Conversion 换算关系
	Name of Unit 单位名称	Symbol 符号	Name of Unit 单位名称	Symbol 符号	
Cubic measure 体积	cubic meter 立方米	m^3	cubic yard 立方码	yd^3	$1\ yd^3 \approx 0.7646\ m^3$
	cubic centimeter 立方厘米	cm^3	cubic foot 立方英尺	ft^3	$1\ ft^3 \approx 0.0283\ m^3$
			cubic inch 立方英寸	in^3	$1\ in^3 \approx 1.6387 \times 10^{-5}\ m^3$
Mass 质量	ton 吨	t	long ton 英吨（长吨）	UKton	$1UKton = 2240lb = 1016.047\ kg$
	kilogram 千克（公斤）	kg	short ton 美吨（短吨）	sh ton, USton	$1USton = 2000lb = 907.185\ kg$
	gram 克	g	pound 磅	lb	$1lb = 16\ oz = 453.592\ g$
	milligram 毫克	mg	ounce 盎司	oz	$1oz = 16\ dr = 28.3495\ g$
			dram 打兰	dr	$1dr = 27.34375\ gr = 1.7718\ g$
			grain 格令	gr	$1\ gr = 1/7000\ lb = 64.799\ mg$
Temperature 温度	Kelvin 开［尔文］	K	Fahrenheit degree 华氏度	℉	表示温度差和温度间隔时：$1K = 1℃ \quad 1℉ = \frac{5}{9}K$ 表示温度数值时：$\frac{T}{K} = \frac{5}{9}\left(\frac{\theta}{℉} + 459.67\right)$ $\frac{t}{℃} = \frac{5}{9}\left(\frac{\theta}{℉} - 32\right)$
	Celsius degree 摄氏度	℃			
Capacity 容积	liter 升	L (l)	(UK) bushel 英液蒲式耳	UKbu	$1UKbu = 4UKpk = 36.36872\ L$
	milliliter 毫升	mL, ml	(UK) peck 英液配克	pk	$1pk = 2\ UKgal = 9.09218\ L$
			(UK) gallon 英加仑	UKgal	$1UKgal = 4\ UKqt = 4.54609\ L$
			(UK) quart 英夸脱	UKqt	$1UKqt = 2\ UKpt = 1.13652\ L$
			(UK) pint 英液品脱	UKpt	$1UKpt = 4\ UKgi = 0.56826\ L$
			(UK) gill 英液及耳	UKgi	$1UKgi = 0.142065\ L$
Density 密度	kilogram per cubic meter 千克每立方米	kg/m^3	pound per cubic foot 磅每立方英尺	lb/ft^3	$1lb/ft^3 = 16.0185\ kg/m^3$
	gram per cubic centimeter 克每立方厘米	g/cm^3	pound per cubic inch 磅每立方英寸	lb/in^3	$1lb/in^3 = 27679.9\ kg/m^3$

Name of Quantity 量的名称	Legal Unit 法定计量单位		Common Non-legal Unit 常见非法定计量单位		Conversion 换算关系
	Name of Unit 单位名称	Symbol 符号	Name of Unit 单位名称	Symbol 符号	
Speed 速率	meter per second 米每秒	m/s	foot per second 英尺每秒	ft/s	1ft/s = 0.3048 m/s
			mile per hour 英里每时	mile/h	1mile/h = 0.44704 m/s
Pressure 压强	Pascal 帕［斯卡］	Pa	dyne per square centimeter 达因每平方厘米	dyn/cm^2	1dyn/cm^2 = 0.1 Pa
			pound per square foot 磅每平方英尺	lb/ft^2	1b/ft^2 = 47.8803 Pa
			pound per square inch 磅每平方英寸	lb/in^2	1lb/in^2 = 6894.76 Pa
			kilogram-force per square centimeter 千克力每平方厘米	kgf/cm^2	1kgf/cm^2 = 98066.5 Pa
Force 力	Newton 牛［顿］	N	dyne 达因	dyn	1dyn = 10^{-5}N
			kilogram force 千克力	kgf	1kgf = 9.80665 N
			pound force 磅力	lbf	1lbf = 4.44822 N

References（参考文献）

[1] HOFFMAN E G. Fundamentals of Tool Design［M］. 2nd ed. Dearborn：Society of Manufacturing Engineers Publications, 1984.

[2]《科技标准术语词典》编辑委员会. 科技标准术语词典：第五卷 机械［M］. 北京：中国标准出版社, 1996.

[3] 西安交通大学外语教研室. 简明英汉科技词典［M］. 北京：商务印书馆, 1983.

[4] 郑易里, 等. 英华大词典［M］. 北京：商务印书馆, 1987.

[5] CAMPBELL J S, Jr. Casting and Forming Processes in Manufacturing［M］. Shanghai：The West Shanghai Books Store, 1950.

[6] ASHBY M FRS, CHARLES J, EVANS A G. Die Casting Metallurgy［M］. London & Frome：Butler & Tanner Ltd. , 1982.

[7] WECK M, BIBRING H. Handbook of Machine Tools［M］. London：Great Britain by Page Brothers Ltd. , 1984.

[8] SMITH W F. Structure and Properties of Engineering Alloys［M］. New York：McGraw-Hill Book Company, 1987.

[9] 叶永昌. 英语要领和难点［M］. 天津：天津大学出版社, 1983.

[10] DALLAS D B. Tool and Manufacturing Engineers Handbook［M］. 3rd ed. New York：McGraw-Hill Book Company, 1976.

[11] DE GARMO E P, BLACK J T, KOHSER R A. Materials and Processes in Manufacturing［M］. 6th ed. New York：Macmillan Publishing Company, 1984.

[12] WAGE H W. Manufacturing Engineering［M］. Dearborn：Society of Manufacturing Engineers Publications, 1994.

[13] LINDBERG R A. Processes and Materials of Manufacture［M］. 3rd ed. Boston：Allyn and Bacon, 1983.

[14] DOYLE L E, KEYSER C A, et al. Manufacturing Processes and Materials for engineers［M］. 3rd ed. London：Englewood Cliffs, 1985.

[15] JAMESON E C. Electrical Discharge Machining Tooling, Methods and Applications［M］. Dearborn：Society of Manufacturing Engineers Publications, 1983.

[16] 焦永和, 等. 工程图学（英文版）［M］. 8 版. 北京：高等教育出版社, 2005.

[17] 夏琴香. 冲压成型工艺及模具设计［M］. 广州：华南理工大学出版社, 2004.

[18] 唐一平. 先进制造技术（英文版）［M］. 北京：机械工业出版社, 2002.

[19] 曾志新, 等. 机械制造技术基础（英文版）［M］. 武汉：武汉理工大学出版社, 2004.

[20] RAO P N. 制造技术——铸造、成形和焊接（英文版·原书第 2 版）［M］. 北京：机械工业出版社, 2003.

[21] RAO P N. 制造技术——金属切削与机床（英文版）［M］. 北京：机械工业出版社, 2003.

[22] 刘伟军, 等. 快速成型技术及应用［M］. 北京：机械工业出版社, 2005.